U. S. History
Through a
PRISM

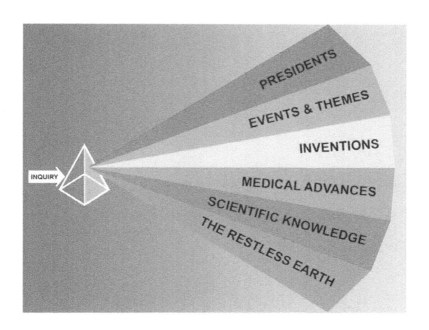

Dr. Michael Slavit

Copywrite © 2022 Michael Slavit

Registration Number: TXu 2-314-828

All rights reserved

ISBN: 979-8-9860153-0-9

Seattle:

Amazon Independent Publishing Platform

Table of Contents

Introduction   page 1

Note about Classification or Taxonomy   page 3

A Note about Eras or Ages   page 4

Era 201   **(1789 – 1817)**   page 7

George Washington   page 7

John Adams   page 10

Thomas Jefferson   page 13

James Madison   page 17

Events that Changed our World   page 19

The Constitution of the United States   page 19

The Louisiana Purchase   page 20

The Lewis and Clark Expedition   page 22

Inventions that would Change our Lives   page 23

The Spectroscope   page 23

The Slater Mill   page 24

The Cotton Gin	p. 24
Medical Advances	p. 25
The Stage is Set for Vaccination	p. 25
First Medical Patents	p. 26
What We Learned about the Universe	p. 26
The Earth is Ancient	p. 26
The Restless Earth	p. 27
The New Madrid Earthquakes of 1811-1812	p. 27
Mt Tambura Volcano 1815	p. 28
Miscellaneous	p. 29
Higher Education	p. 29
Summary Comments on Era 201	p. 29
Era 202 (1817 – 1841)	p. 31
James Monroe	p. 31
John Quincy Adams	p. 34
Andrew Jackson	p. 36
Martin Van Buren	p. 39
William Henry Harrison	p. 43
Technological Advances	p. 44

The McCormack Reaper	p. 44
Volta Invents the Battery	p. 44
The Steam Engine and the Locomotive	p. 45
Medical Advances	p. 46
A Remedy for Malaria	p. 46
What We Learned about the Universe	p. 47
Comet Encke	p. 47
Birth of Paleontology	p. 47
The Molecules of Life	p. 48
The Restless Earth	p. 49
The Great Barbados Hurricane of 1831	p. 49
1840 Natchez Tornado	p. 49
Miscellaneous	p. 50
Growing and Migrating Population	p. 50
Transportation	p. 50
Summary Comments about Era 202	p. 50
Era 203 (1841 – 1861)	p. 52
John Tyler	p. 52
James K. Polk	p. 55

Zachary Taylor	p. 58
Millard Fillmore	p. 59
Franklin Pierce	p. 62
James Buchanan	p. 65
Issues and Themes	p. 68
The Gold Rush	p. 68
The Age of Utopian Thought	p. 69
Inventions	p. 70
The Beginning of Oil Production	p. 70
The Telegraph	p. 71
The Rotary Printing Press	p. 72
Medical Advances	p. 72
Louis Pasteur and the Germ Theory of Disease	p. 72
What we Learned about the Universe	p. 73
The Nature of Scientific Theory	p. 73
The Theory of Evolution	p. 76
The Restless Earth	p. 77
1856 The Last Island Hurricane	p. 77
Miscellaneous Events	p. 78
Seneca Falls Convention and Voting Rights	p. 78

Manifest Destiny	p. 79
Summary Comments on Era 203	p. 80
Era 204 (1861 – 1885)	p. 82
Abraham Lincoln	p. 83
Andrew Johnson	p. 86
Ulysses S. Grant	p. 89
Rutherford B. Hayes	p. 93
James A. Garfield	p. 96
Chester A. Arthur	p. 99
Events and Themes	p. 101
Industrialization and Immigration	p. 101
The Plight of Labor	p. 101
"Robber Barons" or "Captains of Industry"?	p. 102
John Rockefeller and Standard Oil	p. 104
Henry Ford and the Ford Motor Company	p. 105
Andrew Carnegie and Carnegie Steel	p. 106
Inventions	p. 108
The Light Bulb	p. 108
Medical Advances	p. 109

Antiseptics	p. 109
What we Learned about the Universe	p. 110
The Beginnings of Genetics	p. 110
The Restless Earth	p. 111
1871 The St. Louis Tornado	p. 111
1883 Krakatoa	p. 111
Miscellaneous Events	p. 112
Intercontinental Communication	p. 112
Dow Jones	p. 112
Summary Comments on Era 204	p. 113
Era 205 (1885 – 1909)	p. 115
Grover Cleveland	p. 115
Benjamin Harrison	p. 117
Grover Cleveland	p. 119
William McKinley	p. 121
Theodore Roosevelt	p. 122
Issues and Trends	p. 126
The Advent of The Grange	p. 126
Plight of Farmers and Rise of Populist Party	p. 127

The Sherman Antitrust Act	p. 129
Riding the Rails – the Era of the Hobo	p. 130
Inventions	p. 131
The Development of Radio	p. 131
The Wright Brothers' First Flight	p. 132
Medical Advances	p. 133
Walter Reed and the Eradication of Yellow Fever	p. 133
The Discovery of X-Rays	p. 134
What we learned about the Universe	p. 135
The Discovery of Electrons	p. 135
The Discovery of the Atomic Nucleus	p. 136
Einstein and Special Relativity Theory	p. 137
Einstein and General Relativity	p. 137
There was an Age of Dinosaurs	p. 138
Human Origins	p. 139
The Restless Earth	p. 139
The 1906 San Francisco Earthquake	p. 139
The Tunguska Explosion	p. 140
Miscellaneous Events	p. 141
United Fruit Company and American Imperialism	p. 141

Summary Comments on Era 205   p. 142

Era 206 (1909 – 1933)   p. 145

William Howard Taft   p. 145

Woodrow Wilson   p. 149

Warren G. Harding   p. 153

Calvin Coolidge   p. 156

Herbert Hoover   p. 160

Issues and Trends   p. 165

The Alien and Sedition Actis of 1917/1918   p. 165

The Dollar Replaces the Pound   p. 167

The Progressive Movement and Political Labels   p. 167

Women's Right to Vote   p. 171

The Scopes Trial   p. 171

The Birth of the Federal Reserve   p. 172

An Age of Entertainment   p. 174

Prohibition and the Roaring Twenties   p. 175

Break-up of Standard Oil   p. 177

Inventions   p. 178

Sonar   p. 178

Radio Telescope	p. 179
The Ford Model T	p. 180
Medical Advances	p. 181
Treatment of Diabetes with Insulin	p. 181
What we Learned about the Universe	p. 182
Galaxies and the Expanding Universe	p. 182
Human Origins – Discovery of Australopithecus	p. 184
Quantum Theory	p. 184
The Restless Earth	p. 186
1912 Eruption of Novarupta	p. 186
The Mississippi Flood of 1927	p. 187
Miscellaneous Events	p. 187
Refrigeration for Home Use	p. 187
Summary Comments on Era 206	p. 188
Era 207 (1933 – 1953)	p. 192
Franklin Delano Roosevelt	p. 193
Harry S. Truman	p. 197
Inventions	p. 202
The Transistor	p. 202

Medical Advances ... p. 203

Penicillin ... p. 203

The Restless Earth ... p. 204

The Dust Bowl ... p. 204

What we Learned about the Universe ... p. 206

Earth Science – The Earth's Core ... p. 206

Nuclear Fission ... p. 207

Miscellaneous Events ... p. 208

The Great Depression and the New Deals ... p. 208

World War II ... p. 209

The Marshall Plan and the Origin of the Cold War ... p. 210

The Cold War ... p. 211

McCarthyism ... p. 212

Summary Comments on Era 207 ... p. 214

Era 208 (1953 – 1974) ... p. 216

Dwight David Eisenhower ... p. 216

John Fitzgerald Kennedy ... p. 220

Lyndon Baines Johnson ... p. 224

Richard Milhouse Nixon ... p. 229

Events and Themes	p. 235
The Civil Rights Movement	p. 235
Labor Unions Reach their Height of Power	p. 237
Space Exploration	p. 239
The 1969 Moon Landing	p. 239
Inventions	p. 240
The LASER	p. 240
The Origin of the Internet	p. 241
Medical Advances	p. 242
The Polio Vaccine	p. 242
First Organ Transplants	p. 243
What we Learned about the Universe	p. 244
The Ocean Floor	p. 244
First Discovery of a Neutron Star	p. 245
Origin of the Elements	p. 246
Origin of Organic Molecules	p. 247
Confirmation of the "Big Bang" – the CMB	p. 248
The First Identified Black Hole	p. 250
Human Origins	p. 251
The Double Helix of DNA	p. 251

Space Exploration	p. 252
The Launch of the Pioneer 10 and 11 Spacecraft	p. 252
Planet Venus Revealed	p. 253
The Restless Earth	p. 254
Hurricane Carol	p. 254
Miscellaneous Events	p. 255
Alvin and Ocean Floor Exploration	p. 255
The Murchison Meteorite	p. 256
Summary Comments on Era 208	p. 256
Era 209 (1974 – 1993)	p. 258
Gerald Ford	p. 258
James Earl Carter	p. 262
Ronald Reagan	p. 268
George H. W. Bush	p. 274
Events and Trends	p. 278
The Demise of the Fairness Doctrine	p. 278
The Decline in Power of Labor Unions	p. 278
Inventions	p. 279
The First Mobile Phone	p. 279

Medical Advances	p. 280
Magnetic Resonance Imaging	p. 280
Increase in Organ Transplantation	p. 281
What we Learned about the Universe	p. 282
Supermassive Black Holes	p. 282
Investigation of Climate Change and the IPCC	p. 282
Space Exploration	p. 283
Pictures from Mars	p. 283
Voyager 2's Tour of the Outer Planets	p. 283
The Restless Earth	p. 284
Mt. St. Helens	p. 284
Summary Comments on Era 209	p. 286
Era 210 (1983 – 2019)	p. 289
William Jefferson Clinton	p. 289
George Walker Bush	p. 295
Barack H. Obama	p. 303
Donald Trump	p. 309
Joseph Biden	p. 322
Issues and Themes	p. 324

The Rise and Prevalence of Dystopian Literature	p. 324
An Age of Bread and Circuses	p. 325
The Corona Virus Pandemic of 2020	p. 328
Inventions	p. 330
The Hubble Space Telescope	p. 330
Promise of Fusion Power	p. 331
The Discovery and Promise of Graphene	p. 332
Medical Advances	p. 333
Mapping of the Human Genome	p. 333
Antibiotics and Drug Resistant Pathogens	p. 334
What we Learned about the Universe	p. 335
The Kuiper Belt	p. 335
Discovery of Kuiper Belt Objects	p. 336
Discovery of Extrasolar Planets	p. 337
Advances in Origins-of-Life Research	p. 337
Space Exploration in Era 210	p. 339
Roving the Surface of Mars	p. 339
The Galileo Mission to Jupiter	p. 340
The New Horizons Mission to Pluto and Beyond	p. 341
The Restless Earth	p. 341

The 2013 Chelyabinsk Meteor	p. 341
Eruption of Mt. Pinatubo	p. 342
Hurricane Katrina	p. 344
Hurricane Maria	p. 344
Hurricane Harvey	p. 345
Miscellaneous Events	p. 345
Has Foreign Policy of Containment been Effective?	p. 345
Intergovernmental Panel on Climate Change	p. 346
Summary Comments on Era 210	p. 347
Brief Summary of the Journey	p. 354
References	p. 359
About the Author	p. 362

Introduction

This book is about perspective and context. I will provide what I hope is an interesting view of the history of our nation from the presidency of George Washington until the present. I will describe the lives and times of all forty-six U.S. Presidents, but that is not what I hope will make this reading experience enlightening. I have divided the past two hundred and thirty years into ten eras. The eras are arbitrary (this will be explained in the next section). For each of the ten eras I have written sections on the following:

*** Descriptions of the era's Presidents,

*** Issues, events or themes of the era.

*** Inventions that would change our lives.

*** Medical advances that would change our lives.

*** What we learned about the Universe, Earth and life.

*** The restless Earth (earthquakes, volcanoes, hurricanes, tornadoes, exploding meteors, et cetera)

*** Miscellaneous events or themes.

*** Summary Comments on the era.

The descriptions of the Presidents and the events in each of the ten eras are summaries, not comprehensive analyses. There are professional historians who devote their careers to studying the lives, philosophies and actions of one or two Presidents, and the events that took place during their political careers. Writing this book has not rendered me a presidential scholar and reading it will not make you one.

1

As I researched the subject, I found many more events, issues, trends, historical figures, scientific and medical advances, et cetera than I could possibly include. Many events and issues can be and are the subjects of volumes on their own. I have limited the number of topics, and the length of treatment of each topic, in order to keep the book readable.

I am using two metaphors to describe the format of this book. The first, as depicted on the cover, is that of a prism. When we shine the light of inquiry on American history, we can see an array of concepts, just as we see an array of colors when we see light that has gone through a prism. The second metaphor is that of a walk through a forest. The expression "can't see the forest for the trees" implies that when we are too close to a situation, or we are looking at too many details, we may need to step back to get a broader perspective. In a sense, I am trying to present American history in a way that will give a picture of the progression of the entirety of that history without getting lost in the details.

It is easy to push an analogy too far, and I will try to use this one sparingly. Imagine that we will take a two-and-a-half-mile walk through the forest of American history. We will not have time to describe every stand of trees, or to examine each species of tree in detail. The species we will have examined will be those that I have stated: Presidents, inventions, medical advances, knowledge of the Universe, natural events and some miscellaneous species. After each quarter-mile, we will stop, rest and consider what we saw in the previous quarter-mile section. We hope to have an appreciation for that quarter-mile section as a whole. After we have completed our entire two-and-a-half-mile trek, we should have a pretty good understanding of the forest we have seen. We may wish to go back at our leisure for a closer look at some of what we have seen, but we have had a good overall view.

By learning about each era in terms of the events, themes and advances I have chosen, we may gain a better perspective on what life was like during those eras. We may understand each era in a more complete way. Have I selected the most significant issues, events or themes in each era? Probably not, since there are so many ways to select and interpret events. Have I selected the most significant inventions and medical breakthroughs made in each era? Probably. Have I picked out good examples of what we learned about the Universe and the Earth during each era? Probably. Have I described each topic in all its detail? Absolutely not. However, I am confident that the issues, events, themes and advances I have described are significant. It is my hope that this approach will give our nation's history more perspective and texture, and will enable us to better appreciate our history in a broader and more meaningful context.

Note about Classification or Taxonomy

As we move about and perceive our world, we are continually placing objects, people and events into categories. If we did not do this, our life would be impossible. Every moment of our life would be a novel experience and we would have to continually re-invent our understanding of the world. Not being able to classify objects, people and events into categories would be the equivalent of having no memory.

For example, suppose you witness a group of people moving about in a frenzied way. What are you seeing? It could be shoppers at a basement sale scrambling to get the best items. Or, it could be a hockey game, or a violent and dangerous riot. Our ability to quickly recognize and categorize frenzied activity enables us to respond appropriately.

As we have expanded our knowledge of the world, we have invented categories – we have placed objects, processes and periods of time into groups with similar characteristics. Such a system can be called a classification system or taxonomy. The advantage of a taxonomy is that we can quickly gain some understanding of people and events. The problem is that once we have placed a phenomenon into a category, we may believe prematurely that we understand it. Our ability to see people and events in novel ways may be precluded by our assumptions.

A Note about Eras or Ages

Writers have a great tendency to want to categorize people, events and eras. We hear about "The Flower Child Generation" or about "Generation X," or about "Millennials." We hear about the Age of Enlightenment, the Renaissance, the era of exploration, the reconstruction era, et cetera et cetera et cetera. To a degree, these designations are helpful in that they give us snapshots of eras, and a way of differentiating them from other times. However, they may be misleading.

First, they make a tacit assumption that we can paint an era, or the events and persons within it, with the broad sweep of one brush. It should be clear that many, or even most, of the individuals living within a so-called "era" are unaware of or largely untouched by what we, in retrospect, assume broadly characterized the era. For instance, the 1960s are often characterized as "The Flower Child Generation." The image of the 1960s college student questioning the establishment, dropping out of college and living in a commune has become an icon. However, millions of young people in that era went to college and stayed there. Millions went to work after high school. And millions languished in ghettos with neither employment nor education. Most of those many millions of individuals were aware of the flower child idea through stories, news and movies. However, relatively few would say that those images significantly characterized their lives.

In this book I will be grouping numbers of presidential terms together and, alas, I will be grouping them into eras. Even though I will do so for purposes of illustration, please do not look at this as a suggestion that all or most of the individuals living at that time would have felt the presence of that era. Moreover, I have decided against classifying each so-called "era" by a single descriptor. To use a single descriptor, or label, could give a misleading picture, as each era could be classified in so many ways. For instance, think of the era after the end of the civil war. This has been referred to as the "Reconstruction Era," a reference to the reorganization of economic, political and social activity in the South after the Civil War. However, that era could also be labelled as the time when railroads were completed to the west coast, bridging the gap between the coasts and changing our perception of the nation. In addition, the post-civil war years could just as easily be called the era of the birth of commercial oil production. No one descriptor truly captures the essence of an era.

I began by referring to each era simply by a number. As I have re-read my drafts of the book, I found that referring to an era merely by number was not adequate. Therefore, I have added a few descriptors for each era in addition to the number. These descriptors are intended *not* to fully categorize the era, but to help readers re-orient themselves to the era to which I am referring. Therefore, you will see each era entitled by a number plus two or three brief descriptors. There will be TEN eras described, with an average length of twenty-three years.

Using numbers to label an era in American history poses another question. Although this book will begin with the election of George Washington as the nation's first president in 1789, is the number "1" appropriate as a label for that era? Perhaps the number "1" should be used to designate the establishment of the first British colony in Jamestown, Virginia in 1607. In that case, the era beginning with Washington's presidency in 1789 would be era number "9". Perhaps the beginning of the European age of discovery should mark the beginning point in our numbering system. 1492 is often cited as a starting point of that age, which would relegate the era of Washington's presidency to number "14". On the other hand, perhaps the earliest era of history

5

should be that which coincides with the first written records, believed to have occurred in Mesopotamia in 3200 BC. In that case, Washington's presidency would be roughly era 201. If we were to think of the first era as the appearance of the first anatomically modern human beings 300,000 years ago, Washington's presidency would be roughly era number 13,061. I have decided to stick with the number 201, out of respect for the fact that recorded human history has been 5,200 years in the making.

Each era of our history existed in the context of what human beings knew about the Universe, in the context of what inventions and devices had been developed, and in the context of what medical care was available. It is my hope that looking at history with the perspective of the knowledge, endeavors and advancements pertaining to each era will make the nation's history come alive, will give it texture and perspective, and will make it more memorable.

ERA: 201

YEARS: 1789 – 1817

(War of 1812, Louisiana Purchase, Lewis and Clark)

Presidents: Washington, Adams, Jefferson and Madison

George Washington

George Washington lived from 1732 to 1799 and served as our first president from 1789 to 1797. Historians have found it difficult to separate legend from fact regarding President Washington. Washington was born in Virginia, and was the son of a prosperous farmer. His father died when George was eleven years old. As was true for many of our early presidents, George had little formal education. Nonetheless, he loved reading and was self-educated. His self-education is noteworthy in that when an author's ideas resonated with him, he would make those ideas part of his working philosophy. For instance, he read about the rules of civility and decent behavior, and from this learned to be modest in accepting honors. This would be particularly important for U.S. history, as Washington set the tone for a presidency that would not have the trappings of royalty. From *Morals*, by Seneca, Washington learned to control anger. And after reading a drama written by Joseph Addison, Washington became determined to focus on matters of importance and to avoid distraction by inconsequentials.

By age 16 he worked as a surveyor, and by age 21 he had inherited land and was living in comfort as a farmer. He was tall, strong and a good equestrian. He offered his services to Virginia's lieutenant governor, and he had a checkered career in military service to Great Britain.

Washington served for fifteen years in Virginia's legislature. He called it tyranny when the British Parliament was punitive in its response to the Boston tea Party of 1773. He was made a delegate to the Continental Congress in 1774 and he believed armed rebellion was possible. He made known his willingness to lead the colonies' forces, and Congress unanimously chose him to lead the Continental army.

Most historians regard Washington as having been a good general, but not a great one. He lost at least as many battles as he won. But he understood the difficulty Britain faced fighting a war three thousand miles from home. Washington believed that as long as he maintained his army -- in the field and uncaptured – the British could not win. He inspired his men to persist through hardships and, after seven years of fighting, Britain lost its hold on the colonies. Washington returned home to Mount Vernon to resume his life as a farmer.

A defining moment in American history was the 1787 Constitutional Convention in Philadelphia, to which Washington was chosen as a delegate. He had a strong nationalist leaning, and believed in an influential central government. He was unanimously chosen to be president of the convention, and in 1789 he was elected the nation's first President.

As Washington was the country's first President, many of the actions he took did set important precedents. He knew this, and he is credited with this statement: "I walk on untrodden ground. There is scarcely any part of my conduct that may not hereunder be drawn into precedent." Historians credit Washington with several accomplishments. He secured what was then the western frontier with a number of treaties. He established the fiscal structure and currency with which to pay off the national debt.

In 1791 Congress passed Alexander Hamilton's suggested excise tax on spirits. They did not anticipate a vehement response. In 1794 a rebellion broke out in western Pennsylvania, threatening the stability of the young nation. Washington personally led a militia to quell the rebellion. When a war raged in Europe in 1793 after France declared itself a republic and executed its king, Washington maintained American neutrality and avoided a potentially damaging war with Britain.

Washington felt his greatest failure was the split between two of his key advisors, Thomas Jefferson and Alexander Hamilton. By the end of his second term, this schism had resulted in the formation of two separate political parties, something the founding fathers had not anticipated.

George Washington left the presidency voluntarily and after two terms. This was virtually unprecedented in world history and helped set the stage for one of the most remarkable aspects of American democracy: the peaceful transition of government after elections. We may take this for granted and not realize the enormous importance of this legacy of our first President.

Washington left office and returned to his home on Mount Vernon, where he died two years later. Notably, Washington's death from illness was probably hastened by his doctors, who reportedly bled him four times to try to treat an infection. Blood-letting was the most common medical practice at the time. Interestingly, blood-letting remains today the only known remedy for hemochromatosis, or "iron loading disease." But it was a useless and dangerous gesture against all other conditions. It is also a reminder of the primitive state of our medical knowledge, even at a time when brilliant men created a form of government by self-rule that would endure for centuries.

1796			
Presidential Candidate Vice-Presidential Candidate	State	Party	Electoral Votes
John Adams Thomas Pinkney	Massachusetts South Carolina	Federalist	71 58
Thomas Jefferson Aaron Burr	Virginia New York	Democratic-Republican	68 30

John Adams

John Adams lived from 1735 to 1826 and served as the nation's second President from 1797 to 1801. Adams was born in Braintree, Massachusetts. He grew up on his family's farm, and enjoyed his boyhood activities of making boats, playing marbles, wrestling, skating, swimming and shooting. But he was a brilliant individual and the rural life could not satisfy him. He enrolled in Harvard at age 16 and four years later received his degree and went to work as a schoolmaster. He completed legal training at the age of 23 and returned to Braintree to practice law.

Adams was an extremely serious individual dedicated to self-improvement. His thirty-year diary provided insight on his character, and served historians as a source of information on early America. He was deeply introspective in the Puritan tradition. He was relentlessly self-critical and preoccupied with how the human spirit can be corrupted by ambition and wealth.

Historians note that Adams' devotion to personal liberty was ambiguous. At age 26 he listened to the words of an older lawyer, James Otis, who protested some of the repressive methods of the British king. Adams was said to have been "stirred to his deepest being" by the cause of liberty. However, after losing to Thomas Jefferson in his bid for a second term as President, Adams made last minute appointments of judges who were more apt to deter than to support personal liberty.

Adams was a leader during the American Revolution. He was critical of Great Britain's authority in colonial America and viewed British imposition of high taxes and tariffs as tools of oppression. Historians describe Adams as having been intelligent and patriotic, but opinionated and blunt. From 1774 through 1778 he served on the Continental Congress, the new nation's government prior to the adoption of the Constitution. Further, he was influential in the debates that led to the passage of the Declaration of Independence. In the 1780s, Adams served as a diplomat in Europe and helped negotiate the Treaty of Paris (1783), ending the eight-year American Revolutionary War. From 1789 to 1797, Adams was America's first vice president before serving a term as the nation's second president.

France and England were at war. Alexander Hamilton and his followers wanted to strengthen ties with England by joining the war against France. Adams believed, as George Washington had, that it was better to stay out of European affairs. In 1976 French ships began capturing American merchant ships. Adams asked Congress for warships and a navy was established. American and French ships battled with no actual declaration of war.

Events known as the "XYZ affair" unfolded. Adams sought to de-escalate tensions with the French, whose foreign Minister Talleyrand asked for a bribe. Though Adams refused, the French did finally establish peace. Adams was praised for refusing the bribe from the French minister, and avoided a war that might have threatened the survival of the young nation, and considered this to have been his greatest achievement as President.

Adams lost popularity when he signed the Alien and Sedition Acts into law in 1798. This act, written to protect American interests, gave the government broad powers to deport "enemy aliens" and to arrest anyone who strongly disagreed with the government. The Democratic-Republicans led by Thomas Jefferson protested these laws, declaring them unconstitutional. Americans had shed one oppressive government and were fearful their new government could resort to similar tactics. Although the laws were not abused and had built-in expirations, they hurt Adams and may have cost him the election in 1800.

John and his wife Abigail had always wanted a peaceful home life, and returned to their family farm in Quincy, Massachusetts. Adams seldom left home during the next twenty-six years and threw himself into writing. For the rest of his life, Adams wrote political commentary, as well as his autobiography and extensive correspondence. In 1812, Adams and Jefferson reconciled by mail, and the two former political rivals exchanged hundreds of letters prior to their deaths fourteen years later. Adams and Jefferson died on July 4, 1826, the fiftieth anniversary of the signing of the Declaration of Independence.

1800			
Presidential Candidate Vice Presidential Candidate	State	Party	Electoral Votes
Thomas Jefferson Aaron Burr	Virginia New York	Democratic-Republican	73 73
John Adams Charles C. Pinkney	Massachusetts South Carolina	Federalist	65 64

1804			
Presidential candidate Vice presidential candidate	State	Party	Electoral Votes
Thomas Jefferson George Clinton	Virginia New York	Democratic- Republican	162
Charles C. Pinkney Rufus King	South Carolina New York	Federalist	14

Thomas Jefferson

Thomas Jefferson lived from 1743 to 1826 and served as our third president from 1801 to 1809. Jefferson was one of the most brilliant individuals to ever occupy the presidency. When John F. Kennedy was president, he once hosted a group of Nobel Prize winners for lunch at the White House. He is reported to have commented to the assembled Nobel Prize winners, "I think this is the most extraordinary collection of talent and human knowledge that has ever been gathered together at the White House, with the possible exception of when Thomas Jefferson dined alone." Obviously, Kennedy had tremendous admiration for the brilliance of Jefferson.

Jefferson's father, Peter Jefferson, died when Thomas was fourteen years old, leaving his widow and children with land and wealth. At the age of sixteen Jefferson enrolled at the College of William and Mary. He is described as having been quiet and reclusive, and was devoted to reading and violin. He graduated from William and Mary at age nineteen, and then studied and practiced law. Jefferson was regarded as more cautious than innovative as a lawyer. He was not gifted as a public speaker, but was brilliant and had ideas to convey. Thus, he began communicating his ideas more by writing letters and articles than by public speaking.

Before attaining the age of thirty Jefferson had studied many languages as well as mathematics, law, science and philosophy. He excelled in many areas, invented the American system of money and was a designer of homes. At age twenty-six he was elected to the Virginia legislature. He was concerned about the growing tensions between Great Britain and the Colonies. When he wrote an article entitled "The Rights of America," he became well-known throughout the Colonies. The British were so concerned about his ideas that they declared him a traitor.

In 1775 Jefferson was chosen as a delegate to the Continental Congress. He returned to Virginia for a year to care for his ill wife and his daughter. Upon his return to Philadelphia in 1776, he was appointed to a committee to write a "Declaration of the Causes and Necessity for Taking up Arms," and Jefferson was given the task of writing the document. The Congress deleted about a quarter of Jefferson words, including his passages condemning the king for the slave trade. But Thomas Jefferson is the author of the Declaration of Independence.

During the Revolutionary War Jefferson served in the Virginia legislature and later as Virginia's governor. He managed to have a law passed that guaranteed freedom of religion, the first such law in America. Jefferson's bill contained the following: "No man shall be compelled to frequent or support any religious worship, place or ministry whatsoever . . . all men shall be free to profess, and by argument to maintain, their opinions in matters of religion." Separation of church and state is central to freedom in the United States, and we owe a huge debt to Thomas Jefferson in this regard.

After the Revolution, Jefferson served as minister to France. When the United States' Constitution was adopted and George Washington was elected President, Jefferson was appointed Secretary of State. This was a time in which America was inventing itself as a nation. There were philosophical differences among members of Washington's cabinet. Secretary of the treasury Alexander Hamilton believed that a king or a lifetime president would be needed to maintain order. Jefferson, on the other hand, feared dictatorship and believed the people could govern themselves. He also believed that self-government required education, and he believed all children should go to school through third grade (The first mandatory education act would be passed in Massachusetts in 1852).

We must place ourselves in the mindset of a late-eighteenth/early nineteenth century person to appreciate these ideas and conflicts. Though small schoolhouses and some universities did exist, most people had no formal education. The rulers of all western nations were members of royal families. Ascent to power was by succession and not by election. Rule by the people had essentially been absent from the western world since some of the Greek city-states practiced it two thousand years in the past. Self-rule and mandatory education were radical ideas at the time.

Thomas Jefferson was elected President after John Adams' one term, and served two terms. He feared that France, who owned land on both sides of New Orleans, had power to disrupt American commerce by closing the Mississippi to American boats. He thus offered to buy New Orleans. It was a surprise to American envoys when Napoleon offered the entire Louisiana Territory in order to continue to finance his wars in Europe. Although Jefferson believed a President should not assert powers not specified in the Constitution, the opportunity was too good to pass. He signed the treaty buying the Louisiana Territory and later asked Congress for permission for what he had done.

The men who followed Jefferson's ideas were called Republicans. Years later they began to be called Democrats. The men who were aligned with Alexander Hamilton became known as Federalists, and so political parties were born. After Washington's first term in office, Jefferson resigned and returned to Virginia. After Washington's departure from office John Adams, a Federalist, was elected President and Jefferson, a Republican, was elected Vice President. It would be years before election procedures were changed so that candidates for President and Vice President would appear on a single ticket. The men who had written the Constitution had not conceived of political parties. At that time, the candidate with the most votes would be president and the man with the second most votes became Vice President.

Jefferson favored frugality and limited government. In contrast to the opinion of Alexander Hamilton, Jefferson considered the national debt to be harmful. He felt that taxes would favor the privileged classes and hurt the common man. He and his treasury secretary, Albert Gallatin, devised a plan to run the national government on tariffs and to eliminate all internal taxes. The debt was thus cut in half.

In Jefferson's inaugural address he had stated, "Every difference of opinion is not a difference of principle." Thus, he set a tone for the democratic process, and for meaningful discussion of issues. We can promote the same values while debating the best ways to achieve them. Jefferson, like Washington, refused to serve a third term as President.

When his handpicked successor, James Madison, was inaugurated, Jefferson retired to Monticello, his Virginia home. He was happy to leave his executive duties behind and to pursue science and natural history. He continued as president of the American Philosophical Society and corresponded with friends and acquaintances all over the world. His pet project was the University of Virginia. He viewed his work on the university as a fitting conclusion to his life of public service. Jefferson brought the University into being, designing all its campus buildings, setting up its curriculum, and selecting its faculty.

Jefferson enjoyed his popularity until becoming ill in early 1826. Jefferson and John Adams were to be the honored guests on the fiftieth anniversary of the presentation of the Declaration of Independence. However, both former presidents were gravely ill and, perhaps fittingly, they both died on the fourth of July.

1808			
Presidential candidate Vice-Presidential Candidate	State	Party	Electoral Votes
James Madison George Clinton	Virginia New York	Democratic-Republican	122
Charles C. Pinkney Rufus King	South Carolina New York	Federalist	47

1812			
Presidential Candidate Vice-Presidential candidate	State	Party	Electoral Votes
James Madison Elbridge Gerry	Virginia Massachusetts	Democratic-Republican	128
DeWitt Clinton Jared Ingersoll	New York Pennsylvania	Federalist	89

James Madison

James Madison lived from 1751 to 1836 and served as our 4th president from 1808 to 1817. Madison grew up on his family's plantation in Orange County, Virginia. He had fragile health and was schooled at home until age eleven. When his health improved, he attended the College of New Jersey (later renamed Princeton) and was a dedicated student. Discussions at college centered on the trouble between the American Colonies and Great Britain. Madison was described as a voracious reader and he sailed through Princeton in two years. He considered becoming a preacher, but decided instead on politics. Madison was a forward thinker in that he believed religion and politics should be separate. The doctrine of separation of church and state is one of the most important underpinnings of freedom in America today, and James Madison shares credit for this vital principle with his predecessor, Thomas Jefferson.

Madison served on the Continental Congress during the Revolution. After the war he and other leaders called for a convention to form a central government. Madison was a prominent leader at the convention in Philadelphia in the summer of 1787. Madison has been called the "Father of the Constitution" as he led most of the discussions that led to its writing. He was elected to the House of Representatives of the new government, where he led the fight to add the first ten amendments to the Constitution, now known as the *Bill of Rights.*

President Jefferson appointed Madison secretary of state. He and Jefferson worked closely and, when Jefferson retired after an eight-year presidency, Madison was elected President. The autonomy of the new American state was still in doubt, as both British and French ships captured American ships to prevent trade with one another. Young men in Congress known as "War Hawks" talked Madison into asking Congress to declare war on England. This led to the War of 1812, a war that did not go well for the young country. The British captured the city of Washington and burned the Capitol building and the White House, but were turned back at Baltimore. The war was eventually indecisive, and a peace treaty was signed in 1815. News in those days travelled no faster than a ship could sail or a horseman could ride, and the famous Battle of New Orleans was fought weeks after peace had been declared. The Battle of New Orleans was a decisive American victory led by General Andrew Jackson. The War of 1812 did little to solve the issues that led to it. However, it may have done a great deal for American confidence due to Jackson's victory at New Orleans, and due to the young nation having stood up to Great Britain, at that time the most powerful nation on Earth. It also propelled Jackson into prominence as a national figure.

Madison continued to care deeply and communicate publicly about national affairs after retiring to his home in Virginia. He felt the aims of ardent states' rights proponents were a threat to the survival of the Union, and he spoke out against that influence. In a note opened after his death in 1836, he had written, "The advice nearest to my heart and deepest in my convictions is that the Union of the States be cherished and perpetuated."

Events that Changed our World in Era 201

<u>The Constitution of the United States</u>

The Constitution of the United States is a document of immense importance. It set this country on a course of representative government, respect for the rule of law, and protection of individuals from powerful persons in government. Throughout most of recorded human history, individuals have not enjoyed such protections, but have been subject to the power of rulers.

The Magna Carta was an important landmark in the trend toward protection of individual rights. It was drafted by the Archbishop of Canterbury to mediate between the King of England and rebellious barons. It set limitations on the power of the king to dominate the lives of the barons.

Magna Carta is still an important symbol of liberty, and it influenced the writers of the American Constitution. It has been described as "the greatest constitutional document of all time – the foundation of the freedom of the individual against the arbitrary authority of the despot."

After the Revolutionary War, leaders such as James Madison, Alexander Hamilton, and George Washington feared the collapse of their young country. The Articles of Confederation had given the Confederation Congress limited powers and no enforcement authority. The Continental Congress could not print money or regulate commerce. The states were contesting territory, war pensions, taxation, and trade. Alexander Hamilton helped convince Congress to organize a Grand Convention of state delegates to work on revising the Articles of Confederation.

In May of 1787 the Constitutional Convention assembled in Philadelphia. They were so concerned about secrecy that the delegates shuttered the windows of the State House so they could speak freely. They had gathered with the charge to revise the Articles of Confederation, but by mid-June they decided to completely redesign the government, which they did after three hot summer months of heated debate.

James Madison is known as the Father of the Constitution. He spearheaded both its drafting and ratification. Madison also drafted the first 10 amendments -- the Bill of Rights. He had helped develop Virginia's Constitution eleven years earlier, and the "Virginia Plan" was the starting point for debate. In the *Federalist Papers*, Madison wrote, "In framing a government which is to be administered by men over men, the great difficulty is this: You must first enable the government to control the governed; and in the next place, oblige it to control itself."

The framers of the Constitution were obsessed with the concept of liberty, and designed a government without too much power in too few hands. They created three branches: legislative, executive and judicial, with a system of checks and balances to ensure that no one branch could become too powerful. The Constitution became the official framework of the government on June 21, 1788, when New Hampshire became the ninth of thirteen states to ratify it. The U.S. Constitution is one of the longest-lived and most emulated constitutions in the world.

The Louisiana Purchase

The history of control over what has been called the Louisiana Territory is a tortuous one. Beginning in the 17th century, France explored the Mississippi River valley and established scattered settlements. By the mid-eighteenth century, France controlled more of the present-day United States than did any other European power.

In 1762, following the "The Seven Years War" between France and Britain, Spain took control from France of the territory west of the Mississippi. Free transit of the Mississippi to the sea was a vital issue for the United States. A treaty signed with Spain in October, 1795 gave American merchants "right of deposit" in New Orleans, granting them use of the port to store goods for export. Americans used this right to transport products such as flour, tobacco, pork, bacon, lard, feathers, cider, butter, and cheese. Spain also recognized American rights to navigate the entire Mississippi. However, in 1798, Spain revoked the treaty.

Due to Napoleon's strength, Spain ceded the Louisiana territory back to France in 1800. In January 1803 Robert Livingston traveled to Paris to negotiate the purchase of New Orleans from France. His instructions were to purchase New Orleans and its environs. A larger acquisition was not even imagined.

Jefferson sent future U.S. President James Monroe to Paris to aid Livingston in the New Orleans purchase talks. In mid-April 1803, shortly before Monroe's arrival, the French asked a surprised Livingston if the United States were interested in purchasing the entire Louisiana Territory. What was known as the Louisiana Territory stretched from the Mississippi River in the east to the Rocky Mountains in the west and from the Gulf of Mexico in the south to the Canadian border in the north. It is believed that economic problems plus other factors combined to make Napoleon willing to sell the territory. They had failed to put down a slave revolution in Haiti and were facing war with Great Britain.

Negotiations were quick and, at the end of April, the U.S. envoys agreed to pay $11,250,000 and to assume claims of American citizens against France in the amount of $3,750,000. In exchange, the United States acquired the vast Louisiana

Territory. The treaty was signed on May 2 and in October the U.S. Senate ratified the purchase.

The Louisiana Purchase of 1803 brought the United States about 828,000 square miles of territory, thereby doubling the size of the young republic. Part or all of fifteen states were eventually created from the land deal, which is considered one of the most important achievements of Thomas Jefferson's presidency.

The Lewis and Clark Expedition

The Lewis and Clark Expedition spanned May 1804 to September 1806. It was the first American expedition to cross the western portion of the United States. President Thomas Jefferson commissioned the expedition shortly after the Louisiana Purchase in 1803 to explore and map the acquired territory and to find a usable route to the Pacific. Very importantly, Jefferson wanted to establish an American presence before Britain or another European power tried to claim the territory.

The group that embarked on the expedition was known as the Corps of Discovery. The Corps included select Army volunteers under the command of Captain Meriwether Lewis and his friend, Second Lieutenant William Clark. The expedition proceeded west from St. Louis and went down the Ohio River. They then went up the Missouri River, a 2300-mile river originating in the eastern Rocky Mountains, crossed the continental divide and proceeded to the Pacific coast.

The expedition had the secondary goals of studying the area and establishing trade with local Native American tribes. The Expedition covered 8,000 miles and took close to three years. Lewis served as field scientist and recorded botanical, zoological, meteorological, geographic and ethnographic information. The expedition returned to St. Louis with journals, maps and sketches and reported its findings to Jefferson.

The Lewis and Clark expedition was a bold adventure with major goals that were largely accomplished. Curiously, it received little attention. Nineteenth century history books barely mentioned it, even during the United States Centennial in 1876. Lewis and Clark gained new attention around the start of the 20th century. There was a 1904 Louisiana Purchase Exposition in St. Louis and a 1905 Lewis and Clark Centennial Exposition in Portland, Oregon. Although these expositions characterized Lewis and Clark as American pioneers, the story was still considered to be insufficient. It was not until the mid-twentieth century that the expedition was more completely researched and celebrated.

Inventions that Would Change Our Lives in Era 201

<u>The Spectroscope Would Eventually Revolutionize Astronomy</u>

Joseph Ritter von Fraunhofer was a Bavarian physicist and manufacturer of optical lenses. He made an enormous contribution to our study of the Universe when he invented the spectroscope. Visible light that appears white to us is composed of all the wavelengths of light in the visible spectrum. If we shine light through a prism, the white light is separated into its constituent colors and we will observe a rainbow with red, orange, yellow, green, blue, indigo and violet.

Fraunhofer was intrigued by this and in 1814 invented the spectroscope. When Fraunhofer used his spectroscope to study the light from the Sun, he noticed dark lines that crossed the spectrum. His experiments led him to the conclusion that different elements and compounds can both emit and absorb light at certain wavelengths.

Fraunhofer looked at the spectra of Sirius and other stars of similar brightness. He found the spectra of Sirius and other similar stars differed from the sun and from each other. He concluded the lines carry information about the source of light, regardless of distance. The dark fixed lines are now known to be atomic absorption lines that show the presence of elements between the light source and the spectroscope. These lines are still called Fraunhofer lines in his honor. The light emitted by a star can be absorbed by the outer layers of that star or by interstellar gas and dust between the star and our instruments. He thus founded stellar spectroscopy, which continues to this day to be a major tool of astronomers in learning about the Universe.

The Slater Mill

The Industrial Revolution began in England with the establishment of textile mills. In America, the first cotton spinning mill, modeled after England's mills, was the Slater Mill, a complex on the banks of the Blackstone River in Pawtucket, Rhode Island.

The mill's founder, Samuel Slater, had apprenticed in England. After coming to the United States, Slater was hired by Moses Brown of Providence, Rhode Island to produce a set of machines, using water power, to spin cotton yarn. Slater had memorized the plans for water power, as he had not been allowed to take written plans out of England. Construction was completed in 1793 with a dam, waterway, waterwheel, and mill. Slater Mill was designated a National Historic Landmark on November 13, 1966. In December 2014, the mill was added to the newly formed Blackstone River Valley National Historical Park.

The Cotton Gin

Farmers were searching for a way to make cotton farming profitable. When Eli Whitney moved to Georgia in search of work, Catharine Greene funded his work on the first cotton gin, and he completed his invention in 1793. This was a major factor in the economy of the southern states in the pre-Civil War era.

A cotton gin is a machine that quickly separates cotton fibers from their seeds, enabling greater productivity than manual cotton separation. Handheld roller gins had been known to the world since at least 500 CE, but they were too slow to be useful in a growing agrarian economy. Whitney's invention used small wire hooks to pull cotton through a wire screen. Brushes continually removed loose cotton lint to prevent jams. The fibers were then processed into various cotton goods such as linens, while undamaged cotton was used for textiles such as clothing. Seeds may have been used to grow more cotton or to produce cottonseed oil. Whitney's cotton gin revolutionized the industry, though it also led to the growth of slavery in the American South. As the demand for cotton rapidly increased, so did the demand for slave labor to produce it.

Medical Advances in Era 201

The Stage is Set for Vaccination

Edward Jenner was an English physician and scientist who lived from 1749 to 1823. Some observers of the day estimated that 60 percent of the population caught smallpox and 20 percent of the population died of it. Interestingly, milkmaids were usually immune to smallpox. Milkmaids did, however, typically have cowpox, a similar but less virulent illness. These facts were not lost on Jenner, who set out to discover what might be at work. Jenner postulated that the pus in the blisters that milkmaids received from cowpox protected them from smallpox.

On 14 May 1796, Jenner tested his idea. He scraped pus from blisters on the hands of a milkmaid who had cowpox. With that pus he inoculated the eight-year-old son of his gardener. Jenner inoculated the boy in both arms that day, which produced in the boy a fever and some uneasiness, but no full-blown infection. The boy was later challenged with infectious material and again showed no sign of infection. The germ theory of disease would not be established for another sixty years. But Jenner's work set the stage for the germ theory of disease, and with it the potential for disease prevention via vaccination.

<u>First Medical Patents</u>

From a medical point of view, the world of Era 201 (1789 – 1817) was rudimentary. It may be difficult for us to reconcile the primitive state of healthcare with the astonishing wisdom and foresight the founding fathers displayed in creating the Constitution of the United States. The U.S. Constitution, with its amendments and revisions over time via case law, has stood the test of time and has helped to sustain our self-rule in a democratic republic for two hundred and thirty-five years. It is a brilliant piece of work.

In sharp contrast was the first patent issued for a medical device in 1796. It was for a "metallic traction" device whose purpose was to cure pain by "drawing off excessive electrical fluid." The first patent for a medicine was also issued in 1796. It was for "bilious pills," which its inventor claimed would cure yellow fever, dysentery, worms and female complaints. Realistic medical practice was yet to come, but our founding fathers managed without it.

What We Learned about the Universe in Era 201

<u>The Earth is Ancient</u>

In 1802 mathematician John Playfair published a book entitled *Theory of the Earth*. In it he described the theories of James Hutton, who is one of a few investigators to have been termed the "father of modern geology." James Hutton was born on June 3, 1726 in Edinboro, Scotland. Hutton was a chemist, agriculturalist and geologist. Until his time, it was widely believed that the Earth was 6,000 years old, based on the Bishop Usher's calculations of the "begets" in the bible. But that belief would be dispelled by empirical observations. Hutton was a keen observer, and his observations revolutionized our understanding of the Earth and its geological processes.

Playfair joined Hutton on an expedition through the countryside, examining exposed layers of rock wherever they could find them. Based on the features they saw in many locations, Hutton reasoned that there must have been innumerable cycles of seabed deposits, uplift and erosion, followed by sinking of the land for further layers to be deposited. Hutton believed that for these cycles to have occurred, geological forces had to have operated over enormous stretches of time. The idea that the Earth was but six thousand years old was refuted.

The Restless Earth in Era 201

<u>The New Madrid Earthquakes of 1811-1812</u>

A series of huge earthquakes and aftershocks rocked the middle of the county beginning on December 16, 1811 and lasting for a few months. The quakes were centered near New Madrid, then part of the Louisiana Territory and now part of southern Missouri. These quakes were unusual in that they occurred in the middle of the North American crustal plate and not near a plate boundary, as is more often the case.

The 1811–12 New Madrid earthquakes began on December 16, 1811 with an initial earthquake that is now estimated to have been magnitude 7.5–7.9. It was followed by a magnitude 7.4 aftershock on the same day. The New Madrid earthquakes remain the most powerful earthquakes to have hit the contiguous United States east of the Rocky Mountains in recorded history. There were intense sand geysers, and the Mississippi River is reported to have run backward.

There are estimates that these temblors were felt strongly over roughly 50,000 square miles, and moderately across nearly 1 million square miles. By comparison, the 1906 San Francisco earthquake was felt moderately over roughly 6,200 square miles.

The underlying cause of the earthquakes is not fully understood. There have been at least two times in the immensely long geological history of the Earth in which all the continents were grouped together in a single supercontinent. The more recent supercontinent, known as Pangea, began to break up two hundred million years ago, during the early Mesozoic Era. Geologists believe the modern faulting is related to an even more ancient geologic feature, known as the Reelfoot Rift, buried under the Mississippi River alluvial plain. The New Madrid Seismic Zone is made up of reactivated faults that formed when what is now North America began to split or rift apart during the breakup of the supercontinent Rodinia during the Proterozoic Era, about 750 million years ago. For perspective, that was a time before the existence of any chordates (animals with any hard parts). The resulting rift system has remained as a scar or zone of weakness deep underground.

In November 2008, the Federal Emergency Management Agency (FEMA) warned that if a serious earthquake in the New Madrid Seismic Zone were to occur today, it could result in the highest economic losses and loss of life in U.S. history. Such a quake could cause catastrophic damage across eight states. The Earth is still restless – still ridding itself of the heat from its formation. The Earth has not settled into a stable period for the convenience of humankind.

<u>The Mt Tambora volcano of 1815</u>

Mount Tambora is on the island of Sumbawa in present-day Indonesia. This volcano experienced a violent eruption on April 10, 1815. It was one of the most powerful eruptions during recorded history. Volcanic eruptions are often not neat, clean events with clear beginnings and endings. In the case of Mt. Tambora, there were steam-blast eruptions for as long as three years after the blast of April, 1815. The eruption had an impact on global climate, and 1816 has been called "the year without summer." There were crop failures in various parts of the world and, in Washington D.C., people wore coats during the 4th of July celebration.

Miscellaneous Events in Era 201

Higher Education

Many factors would fuel an explosive growth of the United States' wealth, power and prestige in the nineteenth century. Among those factors would be a broad system of colleges and universities. Era 201 was an era in which many colleges and universities were chartered or established. These included Bowdoin College, Charleston College, Davidson Academy, Middlebury College, Ohio University, the State University of New York, The University of Georgia, the University of Maryland, the University of North Carolina, Union University, and Worcester Polytechnic Institute.

Summary Comments about Era 201

This was the era of the birth of the nation. It is important to note that the American Revolution was not something that all colonists had desired. When England enacted oppressive policies against the colonies, most colonists felt that it was their rights as Englishmen that had been violated. Nonetheless, events did lead to an independent nation as declared on July 4, 1776. The importance of the work of Thomas Jefferson and James Madison in writing the Deceleration of Independence and the United States Constitution cannot be overstated.

Britain did not truly believe that the Colonies would maintain their independence, but the War of 1812 solidified it. The Louisiana Purchase then doubled the size of the new nation.

This was not the world we have come to know in the present day. It was a world without fast communication. The Battle of New Orleans was fought by two military forces that did not know the war had ended two weeks earlier. In addition, this was a world without refrigeration, electric lighting, paved roads, public education or mechanical transportation.

It was only during this era that it was discovered that the Earth is ancient. There were doctors, but their practice was rudimentary at best. Medical science was not even in its infancy, as evidenced by the blood-letting method that probably hastened George Washington's death. The germ theory of disease was unknown, though Edward Jenner's experiment with smallpox did set the stage for it. Even the idea of washing one hands to avoid contamination was unknown. The Industrial Revolution came to America with Eli Whitney's invention of the cotton gin and the building of the Slater Mill in Rhode Island. Joseph Fraunhofer's invention of the spectroscope set the stage for later advances in our understanding of the nature of stars. The New Madrid earthquakes, the most severe earthquakes the country has experienced, were a reminder that the Earth is still active.

ERA 202

Years: 1817 – 1841

(The Monroe Doctrine, Political Parties, Trains and Canals)

Presidents: James Monroe, John Quincy Adams, Andrew Jackson, Martin Van Buren, William Henry Harrison

1816			
James Monroe Daniel D. Tompkins	Virginia New York	Democratic-Republican	183
Rufus King John E. Howard	New York Maryland	Federalist	34

1820			
James Monroe Daniel D. Tompkins	Virginia New York	Democratic-Republican	231
(unopposed)			

James Monroe

James Monroe lived from 1758 to 1831 and served as our 5th president from 1817 to 1825. Monroe enrolled in William and Mary College at age sixteen, where he was placed in the upper division due to the proficiency in mathematics and Latin he had acquired in prior schooling at Campbelltown Academy. Two years later, in 1775, fighting erupted in Massachusetts between colonial and British forces and Monroe, caught up in revolutionary fervor, joined the Virginia Infantry. He enrolled as a cadet and was soon a lieutenant. In 1776 he was part of George Washington's famous crossing of the Delaware River in the Battle of Trenton. Monroe led a company in a successful attack on a Hessian position that had two cannons. His men captured the cannon, but Monroe suffered a severe shoulder injury. His valor earned him promotions, but he had no field command. He was unable to raise a regiment, and in 1780 he re-rerolled at William and Mary and studied law under Thomas Jefferson.

In 1782 Monroe was elected to the Virginia legislature and in 1783 he was selected by that body to serve on the Second Continental Congress in Philadelphia. Even after the Constitution was ratified to replace the Articles of Confederation, there was a struggle to determine the relative powers of the federal government and the states. Secretary of the Treasury Alexander Hamilton favored strong federal powers, and was opposed by Congressman James Madison and Secretary of State Thomas Jefferson. Monroe joined Madison and Jefferson and helped create the Democratic-Republican Party to oppose Hamilton's federalist Party.

President George Washington appointed Monroe as Minister to France in 1894. There was much ambiguity at that time as to whether France or Britain were the better ally of the United States. Monroe was favorable toward France and, under political pressure from Federalists, Washington recalled Monroe. Monroe was elected Governor of Virginia in 1799, but returned to France at the request of then President Thomas Jefferson to help negotiate the purchase of Louisiana.

Monroe again served in Virginia as legislator and governor before becoming President Madison's secretary of state. Secretary of State was at that time considered a stepping stone to the presidency. Monroe became Madison's Secretary of War. Even though the War of 1812 was not an unqualified success, with the support of James Madison and Thomas Jefferson, Monroe became the Democratic-Republican candidate for the presidency in 1816. Monroe travelled the country to get a feel for what the people wanted. He was remembered for having been an officer in the Revolution, and for having been wounded in battle. He is said to have been met by cheering crowds. The Federalists were no longer popular and Monroe was elected unopposed. A Boston

Newspaper referred to the era as the "era of good feeling" and, though perhaps not accurate, that label has been used to describe the era of his presidency.

Both the United States and Great Britain were maintaining fleets of warships on the Great Lakes, and the specter of renewed fighting remained. Monroe suggested a mutual reduction of the fleets. England agreed, and historians mark this as the beginning of the cooperative relationships that still exists among the United States, Britain and Canada.

There was an important development during Monroe's second term. Spain had claimed most of South and Central America during its era of conquest, but was losing its hold on the region. Many countries were declaring their independence. In 1823 Spain asked France and other European countries to help them reconquer this part of the world. President Monroe responded by declaring the opposition of the United States to any European country interfering in this hemisphere. He declared that countries within the Western Hemisphere "are henceforth not to be considered subjects for future colonization by any European powers." This became known as The Monroe Doctrine and has been a basic part of American policy to this day.

1824			
John Quincy Adams John C. Calhoun	Massachusetts South Carolina	Democratic-Republican	32%
Andrew Jackson	Tennessee	Democratic-Republican	42% ***
William H. Crawford	Georgia	Democratic-Republican	13%
Henry Clay	Kentucky	Democratic-Republican	13%

*** The twelfth amendment to the Constitution stated that if no candidate received more than fifty percent of electoral votes, the House of Representatives would choose the President from among the three candidates with the most electoral votes. Each of the 24 states cast one vote, for the man favored by most of the state's congressmen. Adams received 13 votes, 7 from the states he had carried in the popular election, 3 from Clay states, and 3 from states that Jackson had carried in the popular election. Jackson received 7 votes in the House and Crawford received 4.

John Quincy Adams

John Quincy Adams lived from 1767 to 1848 and served as our 6th president from 1825 to 1829. He was the son of John Adams, the second U.S. President.

John Quincy received considerable early experience. At age eleven his father John Adams went to France as a diplomat and took him to France. He attended school in France and Germany. His father emphasized to John Quincy that he was expected to commit his life to public service. At age fourteen he went to Russia as secretary to the American minister to Russia. He later worked as a secretary for his father, who was involved in writing the peace treaty at the conclusion of the American Revolution.

John Quincy attended Harvard. He was a serious-minded person who disparaged the drunkenness and vandalism of other students. He also viewed the faculty as being less competent than the ideal. His seriousness and dedication would be both a strength and a weakness. As president, he would not relate well with Congress and failed to get important legislation passed. John Quincy graduated at age twenty. Though he did not like studying law, he did so at the insistence of his parents. He was sent by President Washington to represent the United States in Europe. After his return he was elected to the Senate. Though he did not believe in political parties, he was elected as a Federalist. As he sided on issues more with Republicans than with Federalists, he was not elected to a second term. President

Madison sent John Quincy to Europe in 1814, where he helped to write the peace treaty that ended the War of 1812. The next President, James Monroe, appointed Adams to be secretary of state, and Adams helped frame the Monroe Doctrine.

John Quincy Adams ran for President in 1824 against three other strong candidates, including Andrew Jackson. Jackson won the popular vote, but as no one won a majority, the House of Representatives decided the election, and they chose Adams. Historians describe Adams as very intelligent, but as cold and harsh. He was lonely and not well liked, but valued his dignity. He also highly valued education. He asked Congress for a national university, roads, canals and a naval academy. Although these measures were eventually enacted, it was not under his presidency due to his difficulty with Congress. He was defeated by Andrew Jackson in the presidential election of 1828.

Though hurt by his defeat, he answered the call when the state of Massachusetts elected him to Congress. Some people thought it would be a disgrace to serve in Congress after having been President. Adams did not agree, and he served seventeen years in Congress. He fought against slavery and for civil rights and free speech, and he helped found the Smithsonian Institution. He died in office in 1848.

1828			
Andrew Jackson John C. Calhoun	Tennessee South Carolina	Democratic	56%
John Quincy Adams Richard Rush	Massachusetts Pennsylvania	National Republican	44%

1832			
Andrew Jackson Martin Van Buren	Tennessee New York	Democratic	55%
Henry Clay John Sergeant	Kentucky Pennsylvania	National Republican	42%

Andrew Jackson

Andrew Jackson lived from 1767 to 1845 and served as our 7th president from 1829 to 1837. His father died two weeks before Andrew's birth, and he was born in the rural south, probably in South Carolina. He learned to read and write in a frontier school, but he disliked academics and knew little of science, literature or history. He is said to have been irritable, aggressive, strong-willed, and sometimes a bully.

He joined a Revolutionary Army unit at age fourteen. He was reportedly captured by the British and was cut on the head and hand for refusing to clean a British officer's boots. The incident left him with a hatred of the British and a fierce loyalty to the United States. After the Revolutionary War Jackson moved to Nashville, then a frontier community, and studied law. He had no military command training but was popular. When the people of Tennessee raised an army to fight the Creek Indians, Jackson was elected general. His army was victorious over the Creeks and, in 1814, he was made a general in the federal army.

The slowness of communications in that era was a huge factor in Jackson's career. Although the War of 1812 had ended, news had not reached the British or American armies. Thus, the Battle of New Orleans was fought, and the General Jackson-led American forces dealt the British a crushing defeat. This victory was a major factor in the growing American sense of strength and national pride. Citizens of the Union no longer thought of themselves as recent colonists who might again come under British rule, but as citizens of strong, new nation. And because of the impact of the Battle of New Orleans on national pride and confidence, General Jackson was a national hero.

Jackson was an unusual President in several ways. He was an unrefined man who bet on horses and cockfights. He fought duels, made and lost money repeatedly, and did not actually consider himself to be fit for high political office.

Jackson had had military experience, but no public service experience. As America's political party system developed, Jackson became the leader of the new Democratic Party. He supported states' rights, including the right to own slaves and slavery's extension into the new western territories. He believed in a strict interpretation of the Constitution, and he opposed any government programs that were not explicitly enumerated in it. On that basis he vetoed a bill that would have supported road-building. In his first few months in office, he used the veto power more than all the previous Presidents combined.

Political parties had not been envisioned by the Founding Fathers. They were nonetheless clearly emerging and Jackson may have done much to polarize them. A major battle between the two emerging parties involved the Bank of the United States. The bank's charter was due to expire in 1832. Andrew Jackson and his supporters opposed the bank, seeing it as a privileged institution and as the enemy of the common people. Henry Clay and Daniel Webster argued in Congress for its re-charter. Jackson vetoed the re-charter, charging that the bank constituted the "prostration of our government to the advancement of the few at the expense of the many." Though the National Bank was thought by many to be important to maintain the nation's economic stability, Jackson took independent action and destroyed it before its charter had even run out. Despite the controversial action, Jackson won reelection easily over Clay, with more than 56 percent of the popular vote and five times more electoral votes.

For some, Jackson's legacy is tarnished by his role in the forced relocation of Native American tribes living east of the Mississippi. Although millions of acres of land had been guaranteed to the Cherokee Indians under federal law, the state of Georgia decided to remove the Cherokees. The United States Supreme Court ruled in favor of the Cherokees, stating that Georgia had no authority over Native American tribal lands. But when Georgia acted to expel the Cherokees anyway, Andrew Jackson took no action, stating that Chief Justice Marshall had made the ruling and could enforce it. This led to the infamous Trail of Tears, when in 1838 some 15,000 Cherokee would head on foot to territory west of Arkansas. The relocation resulted in the deaths of thousands. This incident is but one of a number that have inspired some historians to regard Jackson as having been an unlikely symbol of democracy.

During his presidency, Jackson yearned for a quiet retirement. However, he could not let go of politics. Jackson wanted to see his policies carried through and his reputation vindicated. Jackson hand-picked Martin Van Buren to succeed him as President and, during Van Buren's term, Jackson peppered him with advice. He supported Van Buren's Independent Treasury financial plan, intended to stabilize the American financial system by refusing poorly managed state banks access to government funds.

Andrew Jackson had a forceful and dominating personality and a fiery temper he would not always control. Though loyal to his friends, he was not afraid to disregard his superiors, shoot subordinates or captives, or start brawls and street duels. His record is particularly harsh regarding enslaved people and Native Americans. Jackson believed in slavery, and was virulently anti-Indian. He supported the Constitution only when it suited his purposes. Nonetheless he was seen as a symbol of democracy and he received honors during his retirement years. He never tired of praise, jealously guarded his reputation, and spent considerable energy preparing for Amos Kendall's projected biography. Jackson died on June 8, 1845.

1836			elect	pop
Martin Van Buren Richard M. Johnson	New York Kentucky	Democratic	170	51 %
William H. Harrison	Ohio	Anti-Masonic / Whig	73	36 %
Hugh L. White Francis Granger	Tennessee	Whig	26	10 %
Daniel Webster John Tyler	Massachusetts	Whig	14	3%
Willie P. Mangum	North Carolina	Independent / Whig	11	

Martin Van Buren

Martin Van Buren lived from 1782 to 1862 and served as our 8th president from 1837 to 1841. He was born in New York State and had little formal schooling. He worked for a lawyer at 14, became a lawyer at 21 and was interested in politics. He considered himself an intellectual lightweight, but was shrewd, strategic and ambitious. He reshaped American politics, not by pursuing high ideals or inspiring the public, but by promoting political parties and requiring loyalty via patronage.

In 1812 Van Buren ran for the New York state senate, campaigning against the national bank, and was elected. He supported the building of the Erie Canal and, in a progressive move, he sponsored a bill abolishing imprisonment for debt. He served as attorney general as well as being in the state senate, and he used his political influence to secure jobs for persons who would vote for him and for his political allies.

Van Buren was elected to the United States Senate in 1821 and re-elected in 1827. Political parties did exist prior to Van Buren, but they had not been firmly established. In fact, the Federalists had dissolved by the 1820s, leaving the Republicans as the only party standing. Although many Americans, including George Washington and Thomas Jefferson, believed political parties to be evil, Van Buren the strategist saw them as an opportunity.

Van Buren aspired to the presidency himself, but knew he did not have Andrew Jackson's charisma. He therefore chose to work hard for Jackson's election. Some historians view Van Buren as choosing opportunity over principles. He worked hard to create a coalition of northern and southern interests and to create the Democratic Party for the 1828 presidential election. In addition, he resigned from the Senate and ran for governor of New York, thus placing his name alongside Jackson's on the ballot. This tactic worked, and Jackson won New York and the presidency.

When elected, Jackson appointed Van Buren Secretary of State. Van Buren purposefully developed a personal relationship with Jackson. Though he disliked horseback riding, he rode with Jackson to curry favor with him. When Jackson ran for re-election, he selected Van Buren as his vice president. In 1836, with Jackson's support, Van Buren was elected president.

Unfortunately for Van Buren, he became president at an inopportune time. In the previous several years, a speculative fervor had gripped the young nation. Jackson had destroyed the national bank, before its charter had run out, and this action led to an economic collapse. With the steadying influence of the national bank absent, people had been allowed to borrow money to buy land, and to start businesses, in the hope that their speculative ventures would pay off (This would certainly not be the last time in American history that a time of overconfidence and speculation would lead to a major economic downturn). Businesses failed, banks closed their doors, and throughout the country people were unemployed and angry. Van Buren was blamed and, since he had supported Jackson's dismantling of the national bank, he was thus hoisted on his own petard.

Perhaps the best accomplishment of Van Buren's tenure was defusing tensions with Britain. In 1837, Canadian dissidents rebelled against British rule, and many Americans sympathized and helped. This angered Britain, but Van Buren cooled Britain's anger by making it clear that Americans had no right to invade Canada and would have no government support. In addition, there was a dispute between Canadian province New Brunswick and the State of Maine regarding the border, and at stake were 12,000 square miles of valuable timber land. Maine and New Brunswick were arming themselves for conflict, but Van Buren convinced the governor of Maine to withdraw. New Brunswick then showed restraint as well, and armed conflict was avoided.

The slavery issue caused considerable controversy in that era, and Van Buren's preference for expedience over principle was evident. Although the importation of new slaves had been prohibited by law in 1807, the practice continued. Van Buren was equivocal. He condemned the importation of new slaves but, to avoid antagonizing southerners, did little to stop it. In August of 1839 the ship *Amistad* went aground on Long Island and was towed into New London, Connecticut. Africans who had been captured, and who were being sold as slaves, had seized control of the ship and were attempting to return to Africa. Van Buren asked the government to argue in court that the Africans were property and should be returned to their owners. But the court ordered the Africans to be released. Van Buren went so far as to appeal the decision to the Supreme Court, but lost.

In the election of 1840, Van Buren was again hoisted on his own petard. He had helped Andrew Jackson defeat John Quincy Adams by characterizing Adams as a person given to high tastes. These same tactics were used against Van Buren, and he was defeated by William Henry Harrison.

Van Buren stayed politically active after his 1840 defeat. He emerged as a favorite to win the Democratic nomination for the presidency in 1844, but the political landscape was tricky. Whether new states admitted to the Union could be slave states was the controversial issue of the times. Van Buren was equivocal on the issue of the admission of Texas. This angered Jackson, who favored admitting Texas as a slave state, and Jackson threw his support to James Polk, who won the nomination and the election. Van Buren was still not finished politically, as four years later, he became the candidate of a splinter group called the "Free-Soil Party," who called for "free soil, free speech, free labor, and free men." Their main thrust was opposition to slavery in the new western territories. Van Buren had little hope of victory, though his Free-Soil party ran well in several northern states, including New York, Massachusetts, Ohio, and Illinois.

The 1848 election marked the end of Van Buren's active political career. A few years later he wrote his memoir, as well as a milestone study of the organization of American political parties. He traveled, but spent much of his time enjoying his surviving children and grandchildren. Van Buren supported President Lincoln's decision to use force to resist secession from the Union of the southern states. He died at age seventy-nine in 1862.

1840				
William H. Harrison John Tyler	Ohio Virginia	Whig	234	53%
Martin Van Buren Richard M. Johnson	New York Kentucky	Democratic	60	47%

William Henry Harrison

William Henry Harrison lived from 1773 to 1841 and served one month as our 9th president in 1841 until his death due to pneumonia. Harrison was born to a well-to-do family, and grew up primarily in Ohio. He attended Hampden-Sidney College at age fourteen and studied classics. Harrison joined the army after the death of his father, and his career was primarily that of a military man. Harrison's political career began in 1798, when he was appointed Secretary of the Northwest Territory. A year later he was elected to the U.S. House of Representatives. Two years later he became governor of the newly established Indiana Territory, a post he held until 1812.

During his tenure as governor, Indians had been forced off their traditional hunting grounds, and promises to them had been broken repeatedly. In 1811 Shawnee Chief Tecumseh gathered followers and took a stand on the banks of the Tippecanoe River. General Harrison led a force against them and won a victory. Though not a very important battle, it was very significant for Harrison, and it earned him the nickname "Old Tippecanoe."

Harrison served as a general in the war of 1812, after which he moved to Ohio, where in 1816 he was again elected to the House of Representatives. In 1824, he was elected to the U.S. Senate, and left that seat to serve as a diplomat to Gran Colombia, which encompassed much of northern South America. His title there was "Minister Plenipotentiary," indicating he had full power to represent the U.S. government. Believing his career in public life to be over, he returned to private life in North Bend, Ohio. Nonetheless he was nominated as the Whig Party candidate for president in 1836. Though he was defeated by Democratic vice president Martin Van Buren, the party nominated him again in 1840, with John Tyler as his running mate.

The campaign was waged with images and nicknames, with little in the way of issues or philosophy (This would not be the last time the American public would see such a campaign). The Whig campaign slogan was "Tippecanoe and Tyler Too." Although Harrison had grown up in a mansion, it was believed there had once been a log cabin on the property. The Whigs characterized Van Buren as living in a palace (i.e., the White House) and as drinking wine. They declared that military hero Harrison had grown up in a log cabin and drank hard cider. Historians term this the "log cabin and hard cider campaign," and this campaign won the day. They defeated Van Buren in the presidential election, making Harrison the first Whig to win the presidency.

Harrison was worn down by the campaign, by a very long inauguration speech, and by trying to help lines of people seeking assistance. He caught a cold, which developed into pneumonia or typhoid fever, and he died a month after taking office.

Technological Inventions in Era 202

<u>The McCormack Reaper</u>

Until the mid-nineteenth century, grains were harvested either by hand, plucking the ears of grains directly, or grain stalks could be cut with a sickle or a scythe. Robert McCormick designed a reaper in Walnut Grove, Virginia. However, Robert became frustrated with his progress on the device. His son Cyrus asked for permission to complete his father's project. The McCormick Reaper was patented by Cyrus McCormick in 1837. It was a farm implement used to cut small grain crops. The machine was drawn by a team of horses walking at the side of the grain. Although the McCormick reaper was patented in 1837 it did not come into general use until 1845, after which it became a transformative invention in the agricultural sector of the American industrial revolution.

<u>Volta Invents the battery</u>

Alessandro Volta was an Italian physicist and chemist who lived from 1745 to 1827. He held the chair of experimental physics at the University of Pavia for nearly 40 years. He is credited as having invented the electric battery in 1799. Volta proved that electricity could be generated chemically, thereby debunking the prevalent theory that electricity was generated solely by living beings. Volta's invention led to the development of the field of electrochemistry.

There are different kinds of batteries, all functioning via the same basic process. The key is that a battery stores electrical energy in the form of chemical energy and can convert that energy back into electricity. The three main components of a battery are two terminals made of different metals (called the anode and the cathode) and the chemical medium that separates the terminals. The chemical medium is called the electrolyte, and it allows the flow of electrical charge between the cathode and anode. When a device such as light bulb is connected to a battery, chemical reactions occur on the electrodes that create a flow of electrical energy to the device. Volta's discovery was groundbreaking.

The Steam Engine and the Locomotive

As with many technical advances and scientific theories, the invention of steam engines and their productive use was a lengthy process involving many visionaries and innovators. It was seen as a novelty when, in the first century CE, Hero of Alexandria demonstrated that steam could cause objects to move. It was not until the late 17th century that steam was harnessed for practical use. Englishmen Thomas Savery, Thomas Newcomen and James Watt all contributed to the development of steam engines, primarily for use in mining. Englishmen Richard Trevithick and George Stephenson first applied the engine to locomotion by rail.

The first steam locomotive built in America was named "Tom Thumb," and was designed and constructed by Peter Cooper in 1829. Tom Thumb was built as a demonstration to convince owners of the Baltimore & Ohio Railroad that the steam locomotive could be used for commercial purposes.

Railroads created a more interconnected society. The speed and reach of such a dependable mode of transportation changed the day-to-day economics of the nation. People were now able to travel to distant locations much more quickly than if they were using only horse-powered transportation. Counties were able to work together more efficiently due to the decreased travel time. The agricultural productivity of the Midwest could be harnessed once foodstuffs could be shipped inexpensively to distant markets. Trees that thrived in certain areas could be turned into lumber and shipped.

In 1845 Asa Whitney presented a plan to Congress for a transcontinental railroad. Several railways traversing the nation were built in the latter part of the nineteenth century. The first to be finished was the Pacific Railroad, a 3100-mile system completed in 1869. The power and speed of railways led to the development of the coal, iron and steel industries, and contributed to the predominance of the United States as a world industrial power. The steam engine and the steam locomotive were certainly central factors in the transition of America from primarily rural and agrarian to predominantly urban and industrial. They played a central role in the transformation of the nation.

Medical Advances in Era 202

<u>A Remedy for Malaria</u>

Malaria is a virulent disease caused by a microbe — Plasmodium falciparum - carried by and transmitted by mosquitos. Malaria had been endemic in many parts of Europe since ancient times, and there was no known cure. Spanish colonizers learned about an indigenous tree found in Peru. Carolus Linnaeus named the tree 'cinchona,' in honor of a Spanish viceroy's wife who had reportedly been cured from malaria by the plant.

In 1820, French chemists Joseph Caventou and Pierre-Joseph Pelletier isolated the active ingredient quinine from the tree's bark. This made the industrial production of the drug possible. Quinine inhibits nucleic acid synthesis and protein synthesis, thereby impeding the growth of P. falciparum in the patient's red blood cells. Supplies of quinine were particularly important for European armies and colonies, to provide soldiers and colonists a degree of protection from the devastating disease.

What We Learned about the Universe in Era 202

Comet Encke

In 1819 Johann Franz Encke computed the orbit of a dim celestial object first seen by Pierre Méchain in 1786. What became known as Comet Encke is a short-period comet that completes an orbit of the Sun once every 3.3 years. It is somewhat unusual that this object is named after the person who calculated its orbit rather than after the person who first observed it. Most comets and asteroids have what is known as a low albedo; they reflect little of the Sun's light. Comet Encke has a particularly low albedo, reflecting only 4.6 percent of the light it receives. The diameter of the nucleus of Encke's Comet is 4.8 km.

Until the discovery of Comet Encke, Halley's Comet was the only such object known to exist. The discovery of Comet Encke was a signal that the Solar System is a more complex place than had been previously imagined.

Birth of Paleontology

Georges Cuvier was a French naturalist and zoologist, sometimes referred to as the "founding father of paleontology." Although Carolus Linnaeus had traveled extensively and created an elaborate system of classification, his work was mostly descriptive. Cuvier expanded Linnaean taxonomy by grouping classes into phyla and incorporating both fossils and living species into the classification. He was instrumental in establishing the fields of comparative anatomy and paleontology through his work in comparing living animals with fossils.

In that era, it was thought that all living creatures had always existed, and that no previously existing creatures had ever vanished. Cuvier's careful work exploded that myth. Cuvier established that elephant-like bones found in the USA belonged to an extinct animal he later named Mastodon, and that a large skeleton dug up in Paraguay was of Megatherium, a giant, prehistoric ground sloth. These creatures were no longer in evidence. Thus, in his *Essay on the Theory of the Earth*, Cuvier proposed that now-extinct species had been wiped out by periodic catastrophic flooding events. The prevailing theory of the Earth and life was "uniformitarianism" -- the idea that Earth and its creatures change very little, and then only very slowly over time. Cuvier became one of the first proponents of "catastrophism," which is the concept that the Earth and its biosphere undergo periodic, catastrophic changes.

Interestingly, despite his insistence that species do become extinct, Cuvier is known for having opposed theories of evolution. Cuvier could not accept evolution, but believed there had been cyclical creations and destructions of life forms by global extinction events such as deluges.

Cuvier's work is the foundation of vertebrate paleontology, and he is known for establishing extinction as a fact. At the time, extinction was considered by many of Cuvier's contemporaries to be speculation. He described the aquatic reptile Mosasaurus and was one of the first people to suggest that in prehistoric times the Earth had been dominated by reptiles rather than mammals. He collaborated in the study of the strata of the Paris basin and helped establish the basic principles of biostratigraphy. Thus, Cuvier's work gave a boost to our developing understanding of the Earth and the history of its life forms.

The Molecules of Life

Scientists of the era believed that chemistry pertained to rocks and minerals, and that the stuff of life was a totally different domain. A German chemist named Friedrich Wohler dispelled that notion in 1828 when he conducted a series of laboratory experiments. He used cyanic acid, a simple molecule made of carbon, oxygen, hydrogen and nitrogen, and was able to synthesize urea, a component of urine. He had thus demonstrated that the chemicals of life are not different from those of inanimate matter. Wohler's discovery is often cited as having been the starting point of modern organic chemistry.

The Restless Earth in Era 202

1831 The Great Barbados Hurricane

On August 10, 1831 a storm slammed into Barbados, leveling the capital of Bridgetown. This was the Great Barbados Hurricane, an intense Category 4 hurricane that devastated the Caribbean and Louisiana. It is estimated that 1,500 people perished. Many probably drowned in the 17- foot storm surge. But perhaps even more were crushed beneath collapsed buildings. The storm caused great damage in Saint Vincent and Saint Lucia, but slight damage in Martinique.

On August 12, the hurricane reached Puerto Rico, nearly destroying the town of Les Cayes and damaging Santiago de Cuba before crossing the length of Cuba. It passed Havana on August 14. Its fury brought ships ashore at Guantanamo Bay and caused mudslides, resulting in damage to structures.

1840 Natchez Tornado

Shortly before 1 p.m. on May 7, 1840, a massive tornado formed southwest of Natchez, Mississippi. It moved northeast along the Mississippi River, followed the river and stripped forests from both shores. The tornado then struck Natchez Landing, a riverport located below Natchez, tossing 60 flatboats into the river and drowning their crews and passengers. A piece of a steamboat window was found 30 miles from the river. At Natchez Landing, dwellings, stores, steamboats and flatboats were destroyed.

The tornado then devastated the town of Natchez and the nearby Louisiana village of Vidalia. It was reported that "the air was black with whirling eddies of walls, roofs, chimneys and huge timbers from distant ruins . . . all shot through the air as if thrown from a mighty catapult." Forty-eight people were killed on land, and 269 others were killed on the river.

Miscellaneous Events in Era 202

<u>The Growing and Migrating Population</u>

The 1830 census reported a population of 12,866,000, which included two million slaves. There were 10.5 farmers for every city dweller. In 1840 the population had grown thirty-three percent to 17,069,000. There were now 5.5 farmers for every city-dweller. The primary occupation in the country by far was still farming, but industrialization had begun, and with it the migration of people from a rural/agrarian life to an urban/industrial life.

<u>Transportation</u>

Transportation was changed with the building of canals, but the importance of canals was short-lived. The Erie Canal, the longest canal built in the U.S., was opened in 1825. In 1830 in Baltimore, Pierre Cooper designed the first steam locomotive in America. By 1842 Boston and Albany were connected by railway, and there were more than 2,800 miles of railroad in operation in America, twice that of Europe. Thus, canals became obsolete.

Summary Comments about Era 202

This era began after the United States had maintained its independence in the War of 1812. A significant event was the Monroe Doctrine, in which President Monroe declared the era of colonization in the western hemisphere to be over. President Monroe may have also set the stage for growing cooperation among the U.S., Britain and Canada when he negotiated a reduction of navy vessels on the Great Lakes.

Political parties were emerging and becoming distinct and combative. The identification of political parties with certain issues would alternate throughout history. Liberty was always seen as the overriding issue, but opinions on how to preserve it would change. For example, Thomas Jefferson is viewed as on the liberal side of the spectrum, and Alexander Hamilton on the conservative side. Hamilton favored federal control and Jefferson favored states' rights, the opposite of today's view of liberal and conservative. I will argue in my discussion of Era 205 (Telegraph, Populist Movement, Pure Food and Drug Act) that the labels are misleading and meaningless. In Era 202 (Monroe Doctrine, Political Parties, Trains and canals) Andrew Jackson, usually seen as liberal, would hardly be seen that way by today's standards. He favored states' rights, favored slavery in the territories, destroyed the national bank and opposed federal funding of improvements such as road-building.

Era 202 saw the introduction of the McCormack reaper, the machine that would markedly increase agricultural yields. Ironically, this did not turn out to be beneficial to farmers. As we will see in Era 204 (Civil War, First Oil Well, Robber Barons), farmers faced extreme hardship due to several factors, including dropping prices due to increased production. Era 202 saw the beginning of transportation via train and canal, though there were still no paved roads. And, as in Era 201 (War of 1812, Louisiana Purchase, Lewis and Clark), people still managed without electric lighting or refrigeration.

There was an awakening of scientific understanding of the world, as Georges Cuvier demonstrated that species have become extinct, though the theory of evolution was yet to come.

ERA 203

Years: 1841 – 1861

(Seneca Falls Convention, Manifest Destiny, Gold Rush)

Presidents: John Tyler, James Polk, Zachary Taylor, Millard Fillmore, Franklin Pierce and James Buchanan

John Tyler

John Tyler lived from 1790 to 1862 and served as our 10th president from 1841 to 1845. Tyler succeeded President William Henry Harrison when Harrison died a month after taking office. His presidency is unusual in that neither party – Democrats or Whigs – really wanted him to be president, and he was known as "the President without a party." The Whig party wanted to win the election, and Virginia was an important state. The Whigs wanted the southern vote, so they selected Tyler as Harrison's running mate.

Tyler's family was among the elite plantation owners of Virginia. Tyler grew up around politics, as his father was governor of Virginia and a friend of Thomas Jefferson. He graduated from William and Mary College at age seventeen, studied law, and was admitted to the bar at age nineteen. He was well prepared for high political office and served on the Virginia legislature, as governor of Virginia and, in 1816, was elected to the U.S. House of Representatives.

Tyler grew up believing strongly in states' rights. He supported slavery and he believed that the power of the federal government should be strictly limited. He thus opposed federal spending for such projects as roads and harbors. In Congress, his strong belief in states' rights and a strict reading of the Constitution became evident. In 1819, during a war with the Indians, General Andrew Jackson seized Pensacola in what was then Spanish Florida. Tyler denounced Jackson, contending that Jackson had exceeded his constitutional authority.

Tyler was elected governor of Virginia in 1825 and in 1827 was elected to the U.S. Senate. He worked behind the scenes to effect a compromise when South Carolina threatened noncooperation with a federally imposed tariff. President Jackson successfully urged the Senate to pass the "Force Bill," which expanded presidential power and would have allowed Jackson to send federal troops into a state to enforce federal law. Tyler believed fervently that such an expansion of federal power would threaten the Union. His career in the Senate ended when he resigned as a matter of principle. The Senate had previously censured President Jackson for unilaterally dismantling the national bank. Jackson's supporters were now pushing to have the censure expunged. Tyler believed that such a move was unconstitutional. The Virginia state legislature instructed Tyler and fellow Senator Benjamin Leigh to vote for expungement. As he would neither defy the state legislature nor vote for an act he thought unconstitutional, he resigned. Nonetheless, in 1840 the Whig party made Tyler their candidate for vice president with presidential candidate William Henry Harrison.

When Tyler succeeded Harrison, it was felt that he would be a president in name only, and that decision-making would rest with the cabinet Harrison had appointed. Tyler quickly informed the cabinet that, as president, he would be making decisions.

Tyler's presidency was characterized in part by a continuing conflict with Kentucky Senator Henry Clay. Clay aspired to be president in 1844, and he did not want Tyler to be popular and successful. The country was in an economic depression, and Clay and the Whigs wanted a national bank. But when Tyler repeatedly vetoed it, all but one of his cabinet members resigned. Eventually, the Whig party essentially ousted him.

Perhaps the best accomplishment of President Tyler was in the foreign affairs domain. During Van Buren's presidency there was a dispute between Maine and Canada over 12,000 square miles of timber land. Tyler was instrumental in effecting an agreement, and the United States obtained 7,000 of the 12,000 square miles that had been under dispute. Tyler also extended the Monroe Doctrine to include Hawaii. And, in 1844 he signed a treaty with China, granting the U.S. trading privileges.

One important issue during Tyler's presidency was the admission of Texas to the Union. Texas had broken away from Mexico and wanted to be a state. The slavery issue was the predominant issue of the era. It was legal to own slaves in Texas, and the northern states did not want to admit another slave state. The Democrats did want Texas to join the Union, and they won the 1844 presidential election. Congress passed a bill admitting Texas to the Union without waiting for the new President to take office. One of Tyler's last acts in office was to sign the bill admitting Texas. Tyler left office in 1845 and, sixteen years later, was elected to the Confederate Congress. However, he died in 1862, before the congress had assembled.

1844				
James K. Polk George M. Dallas	Tennessee Pennsylvania	Democratic	170	50%
Henry Clay Theodore Frelinghuysen	Kentucky New Jersey	Whig	105	48%
James J. Birney		Liberty	0	2%

James K. Polk

James Polk lived from 1795 to 1849 and served as our 11th president from 1845 to 1849. Polk's presidency was significant in that the borders of the current day continental United States were established.

Polk was born in North Carolina, but his family moved to Tennessee when he was eleven. Polk was described as sickly as a child. He never did attain robust physical health, and was usually in pain. Nonetheless, he possessed what historians would describe as "a fierce, driving spirit." As he was unable to do hard work on the farm or to take part in the rough games of other boys, he spent much time reading. He entered the University of North Carolina at age twenty-one. He excelled at debate and graduated with honors at age twenty-three. He then returned to Tennessee and studied law under a prominent attorney.

For James Polk, law was a route to politics, and he experienced a series of political successes. He served in the U.S. House of Representatives from 1835 to 1839. He worked in favor of laws favored by President Andrew Jackson, and he supported Jackson when he unilaterally withdrew federal funds from the national bank, destroying that institution. America was in a severe economic depression in the late 1830s, and Polk felt his best strategy was to avoid the Washington pollical scene and to return to Tennessee. Though he knew the Tennessee constitution limited the power of the governor, he nonetheless ran for governor, hoping it would be a stepping stone to the vice presidency. He won the governorship, but lost in his bid for re-election - the first defeat in his political career.

Polk tried for the governorship again. The campaign was one of stories and slogans, devoid of issues, and he lost again. He went into seclusion and thought his political career was over.

The 1844 democratic convention was in a stalemate. A major issue was whether Texas was to be admitted to the Union.

Martin Van Buren had angered those in favor of expansion by opposing the annexation of Texas. Van Buren had more than one-half of the votes, but could not attain two-thirds. Andrew Jackson spoke in favor of Polk and, in a compromise move, Polk was nominated. While Polk and the democrats wanted the admission of Texas, the Whig party and their candidate, Henry Clay, did not establish a clear position. Polk campaigned briskly as an expansionist, and won a tightly contested election, partly because third-party candidate James Birney (Liberty Party) drew votes away from Clay.

Polk made it clear from the start that he would be involved in all aspects of his administration. He contended that constant errors would occur if a president entrusted details to subordinates. He had four major goals, all which he attained. He wanted to reduce the tariff due to the hardship it imposed on farmers. He wanted an independent treasury, as he felt state banks were reckless and a national bank favored the wealthy. And he wanted to settle the Oregon boundary issue and to acquire California.

Whigs in Congress passed legislation to fund projects on canals, roads, harbors and rivers, but Polk vetoed every such bill that came to him. He referred to Jackson's veto of a roads bill as a precedent.

Another major issue of the day was the Oregon Territory. All the land west of the Rockies and between California and Alaska was called Oregon, and both the United States and Great Britain claimed it. The 1840's were the iconic years of wagon trains, as many people crossed the country via the Oregon and Santa Fe Trails to establish a new life. A newspaper published a story stating that it was the "manifest destiny" of the United States to stretch from the Atlantic to the Pacific. The phrase became popular. Many people talked about the manifest destiny of the country and, to this day, there are frequently questions about this concept on history tests. The Democratic Party had indicated that the Oregon Territory was worth going to war for if necessary. Neither Polk nor Great Britain wanted a war, so a compromise was reached. An agreement was made that established the U.S./Canadian border that exists today.

President Polk's next territorial target was California. Mexico owned California but had very few settlers there. Polk feared California would be coveted by France or Britain, and he twice offered to purchase California from Mexico and was refused. Having been refused twice, he devised another strategy. He used the southern border of Texas as an excuse to start a fight with Mexico over California. The Nueces River was seen as the border, but Polk now claimed that the Rio Grande was the border, and he sent an army there to establish it. The Mexican government was obliged to fight, giving Polk an excuse to send an army and warships to capture California. The war was more difficult than expected, but ultimately Mexico had to surrender. According to the subsequent treaty of Guadalupe Hidalgo, the United States purchased California and parts of Arizona, New Mexico, Texas, Colorado, Nevada and Utah. Thus, by the end of Polk's presidency, the current borders of the continental U.S. had been established.

When he ran for the presidency, Polk had pledged to be a one-term president. He had never been in robust health, and was exhausted after four years in office. He returned to his home in Nashville, Tennessee and died a few months later.

1848				
Zachary Taylor Millard Fillmore	Kentucky New York	Whig	163	47%
Lewis Cass William O. Butler	Michigan Kentucky	Democratic	127	43%
Martin Van Buren	New York	Free Soil	0	10%

Zachary Taylor

Zachary Taylor lived from 1784 to 1850 and served as our 12th president from 1849 to 1850. He died in office after serving 16 months as president.

Taylor was born in Louisville, Kentucky, which was then a small, frontier town. His early life experiences and interests were military, and he joined the army at age twenty-two. He fought in battles against Indians and in the War of 1812. He was made a general while fighting against the Seminole Indians in Florida. Taylor's career was determined in part by the vision of President James Polk to expand the borders of the country. Polk sent General Taylor to Texas to confront Mexico and to push the southern border south from the Nueces River to the Rio Grande. Historians regard Taylor as having been a good, if not great, military leader.   The advent of the telegraph played a huge roll in the ascension of Zachary Taylor. Americans heard news of the battles against the Mexican army, and Taylor became a national hero when he defeated a larger Mexican army in the Battle of Buena Vista.

When President Polk left office, The Whig party wanted a candidate who would capture voters' imaginations. Two leading candidates were generals: Zachary Taylor and Winfield Scott. They had different styles, as indicated by the nicknames their soldiers had for them. Scott was called "Old Fuss and Feathers," and Taylor was called "Old Rough and Ready." The style of the new frontier nation favored "Old Rough and Ready," and he was nominated by the Whig party and was elected.   He was committed to the survival of the Union, and promised to act independent of party affiliations.

It has been said that Taylor was one of the most uninformed and unprepared candidates to ascend to the Presidency. He had no formal education, read very little throughout his entire life, and spelled poorly in an almost unreadable penmanship. He was honest and blunt, but knew little of issues or politics. It was said he would prefer discussing farming over current events.

The slavery issue was still the most pressing political issue of the day, with special interest on whether western territories would become states and whether slavery would be allowed in them. As Taylor was born in the South and had owned slaves on his plantation in Louisiana, it was generally expected that he would favor slavery. However, his goal was to preserve the Union. Southern congressmen were talking about secession. Taylor warned them that he would lead an army against them if they attempted to break up the Union.

It was during Taylor's presidency that gold nuggets were discovered in the Sacramento Valley in early 1848. This sparked the Gold Rush, one of the best known and most celebrated events in American history of the mid-1800s. The Gold Rush resulted in a huge increase in the population of California, and brought to the fore the issue of California's statehood. Slavery and states' boundaries were among the issues under contention. Whether Texas and/or California would be admitted to the Union as free states or slave states was one issue. In addition, there was a boundary dispute between Texas and New Mexico. Further, there was an issue as to whether runaway slaves had to be returned to the South. Senator Henry Clay advanced compromise bills, but President Taylor vetoed them. Congress worked out a new compromise regarding slavery in new states. Though Taylor may have vetoed it, he became ill on the fourth of July and died five days later.

Millard Fillmore

Millard Fillmore lived from 1800 to 1874 and served as our 13th president from 1850 to 1853. Until age fourteen he worked on his father's small farm in New York's Finger Lakes region. His family was poor, and throughout his life he was motivated to never again sink into poverty. In addition, he was motivated to do what was best for his community and country. Fillmore attended a one-room school and was a willing student. In later years he recalled that, though he learned the alphabet at age seven, he primarily learned "to plow, to hoe, to chop, to log and clear land, to mow, to reap."

At age fourteen he was an apprentice cloth maker and later worked in a mill. He purchased a dictionary and taught himself to read, even while tending the machines. Judge Walter Wood recognized Filmore's potential and advised him to study law. He clerked in a Buffalo law office at age twenty-two and at age twenty-four he was admitted to the bar. He developed a thriving practice and was a leader in his community.

Filmore saw politics as a means both to further ascend from his impoverished past, and to serve the community. Political parties were in a state of flux at that time. Fillmore identified with the aim of the National Republicans to obtain federal funding for improvements, and he also identified with a party known as the Anti-masonic Party, which condemned secret societies. In 1828 he won a seat on the state legislature. In 1832, at age 32, Fillmore was elected to the U.S. House of Representatives. He was involved in the organization of the Whig Party when the National Republicans dissolved two years later.

Fillmore had conflicting desires: to be at home with his wife in Buffalo and to participate in politics in Washington. He left Congress in 1834, but two years later he again won election to the House of Representatives. He opposed President van Buren's plan to establish an independent treasury, and favored a "free banking system," though the means of keeping the system fee of political influence was unclear.

Fillmore represented a district that was strongly abolitionist, and Speaker of the House Henry Clay's slaveholding angered that constituency. Fillmore ran against Clay for speaker and, though he lost, he did become chairman of the Ways and Means

Committee. Though his political career was thriving, in 1842 Fillmore surprised many when he again left politics and returned to Buffalo.

As the 1848 election approached, it seemed the Whigs would make Zachary Taylor their candidate for president. Fillmore is reported to have not had an avid interest in a nomination for Vice President, but accepted it to help the party. Fillmore was elected as Zachary Taylor's Vice President in 1848, and he ascended to the Presidency upon Taylor's death. He would be the last Whig President.

While in Congress, Fillmore had voted against the admission of Texas as a state because Texas allowed slavery. Whether territories admitted as states would be free states or slave states was one of the most hotly contested issues in American history. Fillmore did not favor slavery. However, as was the case for several presidents, preservation of the Union was his overriding priority. Therefore, he supported a law that had been proposed by Senator Henry Clay of Kentucky. Under the law, dubbed the "Compromise of 1850," California was admitted to the Union as a free state. However, it would be illegal to assist a slave in escaping, and slave owners would be allowed to pursue runaway slaves into free states.

Fillmore is credited with an international accomplishment of significance, known as "the opening of Japan." Japan had refused trade with other nations for two centuries. At Fillmore's direction, Commodore Matthew Perry visited the Japanese emperor, who agreed to allow American trading ships access to Japan. In addition, his dedication to the Union and his ability to compromise kept peace, for the time being, among the states.

Millard Fillmore and his family welcomed the escape from Washington after Pierce's election. However, tragedy would mar that escape. First Lady Abigail, having sat outside for hours on Pierce's cold, wet inauguration day, contracted pneumonia and died less than a month later. Soon thereafter, Fillmore's twenty-two-year-old daughter Mary died of cholera.

The former President searched for something to take his mind off these calamities, and politics provided his distraction. Some former Whigs had organized a new party, called the "Know-Nothing Party." This party wished to jealously guard the opportunities of America, and favored immigration restrictions and a waiting period for new citizens to vote. Fillmore refused to promote the anti-immigrant message, but the Know-Nothings and the remaining Whigs made him their presidential candidate in 1856. The former President received twenty-one percent of the vote. Fillmore opposed candidate John Fremont of the new Republican Party, and his twenty-one percent prevented Fremont from winning against Democrat James Buchanan.

Fillmore made no more attempts at political office. He retired in Buffalo, married and was active in causes and charities. He strongly supported the Union during the Civil War and helped with enlistment and war-financing. Fillmore died of a stroke in March 1874.

1852				
Franklin Pierce William R. King	New Hampshire Alabama	Democratic	254	51%
Winfield Scott Wm Graham	New Jersey North Carolina	Whig	42	44%

Franklin Pierce

Franklin Pierce lived from 1804 to 1869 and served as our 14th president from 1853 to 1857. Pierce grew up in a frontier community in New Hampshire. His father, Benjamin Pierce, fought in the Revolutionary War, was a local sheriff and then became governor of the state. Thus, his family went from modest means to being a well-positioned family. He graduated from Bowdoin College in 1824 and studied law. In 1827 he was admitted to the bar. Two years later, and during his father's term as governor, Franklin was elected to the state legislature. He was subsequently elected to the U.S. House of Representatives in 1832.

Pierce and his father were strong supporters of Andrew Jackson's policies. While in the House of Representatives, Pierce sympathized with the southern states and spoke out against abolishing slavery. He is said to have misrepresented the strength of the abolitionist movement in his home state. He strongly disliked abolitionists, and portrayed them as "women and children." Nonetheless, in 1836 the New Hampshire legislature elected Pierce to the U.S. Senate. He was known to be a gifted orator. However, he has been portrayed as having otherwise had modest talent. Further, he enjoyed parties and was a heavy drinker. He became abstinent for a while. He may have returned to alcohol while President, and certainly did in his later years.

Family life was not easy for Pierce, as his wife's health was fragile and two of their three sons died at young ages. Pierce left the senate and returned to New Hampshire at his wife's request.

Pierce enlisted as a private in the army in 1846 when the Mexican War began. Despite his lack of military experience, President Polk promoted him to Colonel and then to General. He returned from the war without the glory to which he had aspired. There was no clearly dominant candidate for President when the 1852 Democratic Party's convention began. The front runners were Stephen Douglas of Illinois, James Buchanan of Pennsylvania and Lewis Cass of Michigan. The delegates were unable on many votes to reach a conclusion. Finally, Pierce's supporters argued that he would unite the party. He was named on the forty-ninth ballot as a compromise candidate, and he won the presidency.

Pierce's first year in office went poorly. On their way to the inauguration, the Pierces' train overturned and their eleven-year-old son was killed. Pierce may never have recovered from the tragedy and his wife became reclusive and is reported to have spent much time writing to her deceased son.

Pierce wanted to expand the United States' territory. He was not very successful in that regard. He failed to acquire Cuba from Spain, or to take over Hawaii, which was independent. In a transaction called The Gadsden Purchase, he purchased from Mexico what is now southern Arizona and the southwestern corner of New Mexico.

Throughout the nation people were still hotly debating the slavery issue. Pierce may have unwittingly fanned the flames of rivalry between north and south due to his actions on this issue. The Missouri Compromise of 1820 had made slavery illegal north of a line. Senator Stephen Douglas, motivated by self-interest, proposed a new law called the Kansas-Nebraska Act. The new act would allow each territory to decide on its own about slavery, and Pierce helped get the new act passed. Pierce was no longer popular, even in his home state, and the Democratic Party did not choose him to run for a second term.

Pierce settled in New Hampshire after his presidency. When the Civil War erupted, he was among those who believed the Constitution protected the institution of slavery and he supported the Southern position. Pierce publicly blamed Lincoln for the war, and his outspoken criticism cost him several friendships.

Pierce had always been fond of liquor, and he returned to it. When Lincoln was assassinated in April 1865, an angry mob surrounded Pierce's home. Pierce spoke to the crowd and asked them to disperse peacefully. They did so, perhaps a tribute to his once-famed skills as an orator.

Pierce is considered to have been a failure as President. The country steadily broke apart over a few decades, and the presidents and politicians who were in power during those years have not been looked upon kindly by historians. Pierce died in the fall of 1869, and relatively little has been written about him.

1856				
James Buchanan John C. Breckinridge	Pennsylvania Kentucky	Democratic	174	43%
John C. Fremont William L. Dayton	California New Jersey	Republican	114	33%
Millard Fillmore Andrew Jackson Donelson	New York Tennessee	American / Whig	8	22%

James Buchanan

James Buchanan lived from 1791 to 1868 and served as our 15th president from 1857 to 1861. He was born in a one-room log cabin in rural Pennsylvania. When James was six, his father moved the family to two-room brick house that served as a home and a trading post. As a boy, Buchanan worked in his father's store, learning arithmetic and bookkeeping. He attended the local Old Stone Academy, studied Latin and Greek, and enrolled at Dickinson College. He attended to his studies at first, but to be popular he started drinking and smoking and, in his own later words, "engaged in every sort of extravagance and mischief." He was expelled, later re-admitted, and graduated in 1809.

Buchanan then studied law and began a practice in Lancaster in 1812. He was elected to the state legislature in 1814, then retired after two terms to concentrate on his law practice. He had considerable wealth by age 27, at which time he proposed marriage to Ann Coleman, who accepted. But her parents disapproved, and Ann broke off the engagement and later died. There were rumors her death was a suicide, and Buchanan blamed himself and never married.

Buchanan returned to politics and was elected to the U.S. House of Representatives in 1820. The Federalist Party was dissolving and Buchanan promoted an alliance of Federalists and Jacksonian Democrats. He was devoted to obstructing the goals of President John Quincy Adams to set the stage for an Andrew Jackson presidency. Historians regard Buchanan as having been blatantly opportunistic, not standing on principle, but rather using the prevailing political tides to advance his own career.

Secretary of State was a typical stepping stone to the Presidency, and President Polk appointed Buchanan to that position. He worked on issues such as the Mexican War and the Oregon Territory. The hottest issue in the country in the 1856 election year was slavery. Southern slave owners favored slavery in all the territories, while in the North the total abolition of slavery was a prevailing view. The new Republican Party was clear in its opposition to slavery. In an apparent attempt not to alienate anyone, the 1856 Democratic Party avoided taking a strong stand, and Buchanan's position was unknown. He was chosen as the Democratic candidate for President and he won.

Buchanan's position soon became clear when the Supreme Court ruled on the case of Dred Scott. Scott was the slave of an army physician in the Wisconsin Territory, where slavery was illegal. When Scott returned to Missouri, he sued for his freedom, citing that he had lived in a free territory, and his case went to the U.S. Supreme Court. Slavery remained the hottest issue of the times, and the Supreme Court decision would affect not just Dred Scott and his family, but also Congress' role regarding the spread of slavery. Buchanan feared the possibility of the South seceding from the Union, and he thus adopted a pro-slavery stance.

In secret communications with two justices of the Court, President-elect Buchanan suggested the Supreme Court rule against Dred Scott. Though Buchanan may have personally opposed slavery on moral grounds, he more strongly wanted to avoid controversy and to further his own political career. The Court's decision would not be officially announced until March 6, 1857, but in his inaugural address of March 4, Buchanan endorsed the Court's decision.

In its 7-2 ruling against Dred Scott, the Supreme Court held that Congress had no right to prohibit slavery in the territories, slaves were property, and slave owners could not be deprived of their property without due process. The Court thus supported the idea that a person could be deemed "property" and had no rights of citizenship. Buchanan urged the country to respect the Supreme Court's ruling. President Buchanan's position, while consistent with the majority in Congress and the Supreme Court, was inconsistent with the nation's highest principles.

Tensions over slavery continued to grow, and this was the prevailing issue as the 1860 election drew near. The Democratic Party was still divided over the slavery issue. Abraham Lincoln was nominated by the Republican Party, and he won.

Southern states were incensed by Lincoln's election to the presidency. They feared that Lincoln would abolish slavery and, in so doing, "end southern civilization." During Buchanan's last few months in office southern states began seceding from the Union. Buchanan apparently did nothing about the issue of secession other than hope that the southern states would rejoin. Lincoln took office on March 4, 1861 and Buchanan finally took a stand and stated that the North must back President Lincoln.

Potentially, a comfortable private life awaited James Buchanan at his Pennsylvania home. However, his last years were difficult and he was blame for the Civil War. Attitudes toward Buchanan were so negative that his portrait had to be removed from the Capitol to prevent vandals from damaging it.

Although Buchanan vocally supported the cause of preserving the Union, many saw him as a proponent of slavery who tried to appease the South. The former President blamed the Civil War on the Republican Party and abolitionists, and he finally decided to write a book advancing that position. It appeared in 1866, one year after the war ended. The public largely ignored the book and Buchanan vanished from public life. He stayed within the walls of his home and saw only close friends. He died there in June 1868.

Issues or Themes in Era 203

The Gold Rush

On January 24, 1848, James Wilson Marshall, a carpenter from New Jersey, was working in the Sacramento Valley. His task was to build a water-powered sawmill for John Sutter, a Swiss citizen and founder of the colony of New Switzerland (The colony would later become the city of Sacramento).

Marshall found flakes of gold in the American River at the base of the Sierra Nevada Mountains. As Marshall later recalled, "It made my heart thump, for I was certain it was gold." Marshall and Sutter tried to keep news of the discovery secret. But word got out, and by mid-March at least one newspaper was reporting the discovery. Of course, this sparked the Gold Rush, one of the best known and most celebrated events of American history.

Throughout 1849, people around the United States (mostly men) did all they could to join the rush, drawn by the promise of wealth they had never imagined. They borrowed money, mortgaged their property or spent their life savings to make the long, difficult journey to California. They left their families and hometowns, leaving women to assume new responsibilities such as running farms or businesses, or caring for their children alone. Tens of thousands of gold-mining hopefuls were known as "Forty-niners." Some traveled overland across the mountains. Some sailed to Panama or even around Cape Horn, the southernmost point of South America, in their efforts to reach California and its promise of fortune.

By the end of 1849, the non-native population of the California territory was estimated to have risen from less than 1,000 to 100,000. A few hundred million dollars' worth of the precious metal was extracted from the area during the Gold Rush, which peaked in 1852.

California's admission to the Union as the 31st state was hastened by the Gold Rush. In late 1849, California applied to enter the Union with a constitution preventing slavery. This provoked a crisis, as whether newly admitted states could permit slavery was one of the biggest issues of the era. Kentucky's Senator Henry Clay made a proposal that became known as the Compromise of 1850. California was allowed to enter as a free state, while the territories of Utah and New Mexico were left open to decide the question for themselves.

The Age of Utopian Thought

The term "Utopia" comes from the title of a work by Sir Thomas More in the sixteenth century. The word "Utopia" derives from the Greek "ou topas," meaning "nowhere." Thomas More's *Utopia* is a series of conversations between fictitious characters: traveler Thomas Morus, and Raphael Hythloday, a member of a society that had some ideal characteristics. For our purposes, the specifics of Thomas More's book are not important. What is critical is that More's work inspired writers in subsequent centuries to devise their own ideas about ideal societies, which were termed "Utopias," or described by the adjective "utopian."

This movement did more than just inspire writers. It also inspired people to plan and attempt to actualize ideal communities. In the United States, there were about one hundred planned utopian communities between 1825 and 1860. These communities were intended to provide an alternative to nineteenth century life. (Whether you lived in a rural or urban setting, nineteenth century life was beginning to be more complex and demanding than life in prior eras).

There is a reason I am highlighting utopian writings and planned utopian communities as a notable theme in this era of American history. There is a German term "zeitgeist," which translates approximately to "spirit of the times." I do not believe that there is any one spirit of the times in any given era. As I explained in the introduction, I do not believe any era can be described by a single term. However, there may be two or three major themes that weave together to form the fabric of a given era. In the mid-nineteenth century, I believe there was a theme of idealistic thinking – a belief in the perfectibility of humankind. Many persons who held such beliefs appear to have felt that their beliefs could more easily find expression in communities excerpted from the mainstream of American life than by participation in the social and economic life of the times. But the important point is that there were writers, community planners and their followers, and many people in the general public who held an optimistic view of human beings and of the potential for better, more ideal living. This contrasts with later eras, when dystopian ideas became more prevalent.

Inventions in Era 203

The Beginning of Oil Production

Prior to 1859, some wells were drilled to extract salt brine, and some of these wells produced oil and gas as accidental byproducts. But the first well whose primary purpose was oil production was that drilled by George Bissell and Edwin L. Drake on August 28, 1859 at a site near Titusville, Pennsylvania. The importance of the Drake Well is not merely the fact of its having been the first well to produce oil. Historians have noted that this event led to the first great wave of investment in the drilling, refining, and marketing of oil. Furthermore, all the enormous economic growth in the world during the past century and a half has been made possible primarily by one cause: the availability of energy. Fossil fuels have provided well over ninety percent of the world's energy over this time period. Due to its significance as a landmark event, the Drake Well was listed on the National Register of Historic Places and was designated a National Historic Landmark in 1966.

The Telegraph

Until the mid-nineteenth century, communication was extremely slow. Essentially, it was no faster than the horse. Recall that the War of 1812 had ended more than two weeks before the Battle of New Orleans, but neither the British nor American armies knew it. The telegraph would be the first significantly faster mode of communication.

Samuel Morse was born in 1791 in Charlestown, Massachusetts and attended Yale University. Though he had an interest in electricity, he became a painter. While sailing home from Europe in 1832 he heard about the newly discovered electromagnet, and he came up with an idea for an electric telegraph.

Morse spent several years developing a prototype. During that time, he took on two partners, Leonard Gale and Alfred Vail, to help him. Morse's device used electric impulses to transmit encoded messages over a wire. On January 6, 1838, Samuel Morse used Morse code and demonstrated his telegraph system in Morristown, New Jersey.

It took until 1843 for Morse to convince Congress to fund the construction of a telegraph line from Washington, D.C. to Baltimore, Maryland. In May 1844, Morse sent the first official telegram over the line. The telegraph would revolutionize long-distance communication, and it reached its height of popularity in the 1920s and 1930s.

Private companies, using Morse's patent, set up telegraph lines around the Northeast. In 1851, the New York and Mississippi Valley Printing Telegraph Company was founded; it later changed its name to Western Union. In 1861, Western Union completed a line across the United States. Five years later, the first successful permanent line across the Atlantic Ocean was constructed and by the end of the century telegraph systems were in place in Africa, Asia and Australia. Although the advent of radio would eventually render the telegraph obsolete, the telegraph did speed up communication for decades.

The Rotary Printing Press

Newspapers were in operation, but their printing methods were slow until Richard Hoe, an American inventor from New York City, designed a rotary printing press in 1843. Hoe's invention used a continuous roll of paper and revolutionized newspaper publishing. It was superior to the flatbed printing process and enabled the printing of 8,000 newspapers per hour.

Hoe continued to innovate, and in 1870 developed a rotary press that printed both sides of a page in one operation. Hoe's press fed eight hundred feet of paper per minute from a roll of paper five miles long. As the printed paper emerged, it passed over a knife that cut pages apart. There was even an apparatus to fold the pages for delivery. This process produced 18,000 papers an hour and was used the first time by the New York Tribune.

Medical Advances in Era 203

Louis Pasteur and the Germ Theory of Disease

Louis Pasteur was a French biologist, microbiologist and chemist renowned for his discoveries of the principles of vaccination, microbial fermentation and pasteurization. Pasteur was born in France in 1822. He was an average student in his early years, and his interests were fishing and sketching. He earned his Bachelor of Letters degree in 1840 and eventually attained his Master of Science degree.

In 1848, he became professor of chemistry at the University of Strasbourg. Pasteur was responsible for disproving the doctrine of spontaneous generation. He demonstrated experimentally that in sterilized and sealed flasks nothing ever developed, and in sterilized but open flasks microorganisms could grow.

He is remembered for his remarkable breakthroughs in the causes and prevention of diseases, and his discoveries have saved many lives ever since. He reduced mortality from puerperal fever and created the first vaccines for rabies and anthrax. His medical discoveries provided direct support for the previously unproven germ theory of disease and its application in clinical medicine. He is best known to the general public for his invention of the technique of treating milk and wine to stop bacterial contamination, a process now called pasteurization. He is popularly known as the "father of microbiology." Today, he is often regarded as one of the fathers of germ theory. He was thus a prime figure in the early development of what would become modern medical practice as we know it.

What We Learned about the Universe in Era 203

The Nature of Scientific Theory

The next section is headed:" The Theory of Evolution." First, I want to make some comments about the nature of a scientific theory.

The word "theory" has a meaning in common parlance, and it has a completely different scientific meaning. In common speech, a "theory" can mean a speculative, fanciful idea with no evidence. In common speech you may say, "I have a theory that we do not exist, but that we are thoughts in the mind of a gigantic interstellar octopus." We could never test such an idea. Wild ideas and fanciful speculations can be great fun, but they are not theories in the scientific sense.

In science, we start with a hypothesis. A hypothesis is an explanation of events that is supported by at least some evidence and is a starting point for further investigation. Once a hypothesis is generated, it must be subjected to investigation, which may mean observation of natural events or controlled experiments. If the evidence does not support the hypothesis, then the hypothesis is dropped, or new experiments or observations are planned.

When a series of observations and/or experiments consistently support a hypothesis, a tentative theory is born. Even then, results are published in scientific papers, and other researchers usually try similar experiments or make similar observations. That is the scientific method. When many observations or experiments support the tentative theory, a full-fledged scientific theory is in place.

Example: Einstein's General Theory of Relativity

In the early part of the twentieth century, Albert Einstein hypothesized that gravity is not a force per se, but that it is a result of the bending of four-dimensional space-time by massive objects. It took years for the evidence to support his idea, but the evidence is now impressive. Identical, coordinated atomic clocks on the Earth and in a plane flying at high altitude will tell time differently, as predicted by general relativity. During a Solar Eclipse, starlight coming from stars behind the Sun will be slightly bent toward the Sun, as predicted by general relativity. The orbit of the planet Mercury does not behave as Newtonian theory predicted, but it is consistent with general relativity. And, our global positioning technology would not work accurately if we did not make adjustments according to general relativity. Einstein's general theory of relativity is not "just a theory." It is at least part of a correct description of reality.

Another example: The Theory of the Greenhouse Effect

Joseph Fourier first argued in favor of the greenhouse effect in 1824. The Earth receives a gigantic amount of solar radiation, much of it in wavelengths of infrared and visible light. Much of this radiation is reflected off the atmosphere and off the Earth's surface - particularly off ice sheets and light-colored desert sands. It took many decades for this process to be well understood.

We now know that greenhouse gases – particularly water vapor, carbon dioxide and methane – stop some of the reflected radiation from escaping into space. These gases thus serve to warm the Earth. Controlled experiments to test this hypothesis were, of course, not plausible. But the Solar System has conducted a natural experiment for us. The planet Mars, with 1/100th the atmospheric density of Earth, is about as cold as it would naturally be. The planet Venus, with an atmosphere 100 times as dense as Earth's has a surface temperature of about nine hundred degrees Fahrenheit – much hotter than it would be without an atmosphere. Earth is about sixty Fahrenheit degrees warmer than it would be without an atmosphere.

In addition, scientists can deduce the history of Earth's approximate temperature and its atmospheric content of greenhouse gases by studying evidence such as ice cores from Greenland and Antarctica. The evidence is strong. The idea that Earth's temperature is influenced by the content in its atmosphere of certain gases is not "just a theory." It is a good approximation of truth.

Conclusion: In common speech, you may refer to fanciful ideas and speculations as "theories." However, it would be better to reserve the word "theory" for scientific ideas that have been tested and re-tested and are considered at least a part of truth. Therefore, when you hear someone say that General Relativity, or the Greenhouse Effect, or Evolution or Plate Tectonics are "just theories," you may want to caution them. In science, "theory" is a strong word referring to ideas that have been tested and supported. Fanciful ideas and speculations can be great fun, but do not confuse them with scientific theories.

The Theory of Evolution

Charles Robert Darwin lived from 1809 to 1882. Darwin has been described as one of the most influential figures in human history. He was an English naturalist, geologist and biologist, best known for his contributions to the science of evolution. Darwin's early training was primarily in geology, and he was a student of the eminent geologist Charles Lyell. His early observations and theories supported Charles Lyell's uniformitarian ideas.

Darwin's five-year voyage on HMS Beagle established him as an eminent geologist. His publication of his journal of the voyage made him famous. However, he was puzzled by the geographical distribution of wildlife and fossils he collected. He began detailed investigations, and in 1838 conceived his theory of natural selection. Although he discussed his ideas with several naturalists, he needed time for extensive research and his geological work had priority. He was writing up his theory in 1858 when Alfred Russel Wallace sent him an essay that described the same idea, prompting an immediate joint publication of both their theories.

In his joint publication with Wallace, Darwin introduced his theory that the branching pattern of evolution resulted from a process he called "natural selection." It had long been obvious to breeders of hunting dogs that selective breeding would change the characteristics of a species. Darwin showed that the struggle for existence has an effect similar to the artificial selection used by dog breeders. He established that all species of life have descended over time from common ancestors. Darwin's scientific discovery is the unifying theory of the life sciences, helping to explain the diversity of life on Earth.

Darwin published his theory of evolution with compelling evidence in his 1859 book *On the Origin of Species*. By the 1870s, the scientific community and a majority of the educated public had accepted evolution. Although many people favored competing explanations, a broad consensus has developed in which natural selection is established as the basic mechanism of evolution. Subsequent scientific investigation has revealed at least a 3.4-billion-year history of life, in which species appear, evolve, and become extinct in life's ever-changing struggle for survival. The phenomenon of extinction was a novel concept at that time. Darwin's painstaking, detailed observations and his brilliant deductions have played a major part in humankind's understanding of science and of life.

The Restless Earth in Era 203

1856 The Last Island Hurricane

1856 was the year of the Last Island hurricane, which is also known as" The Great Storm of 1856." The hurricane was one of Louisiana's deadliest known storms. It produced a surge of 11-12 feet and resulted in at least 200 fatalities. Most of the fatalities were offshore, as at least 183 people drowned when steamers and schooners sank in rough seas.

The ship Star regularly provided service to the mainland, and many vacationers hoping to escape awaited its scheduled arrival. However, the Star was blown off course, and barely escaped sailing into the open gulf, directly into the hurricane, where it would have almost certainly been lost. The captain managed to pull into the channel behind the hotel. The Star was swept into shore and was beached on the sand, where it stayed through the storm.

The storm surge completely submerged Last Island, destroying almost every structure, including hotels and casinos. More than thirteen inches of rain fell in New Orleans, and rice fields were under several feet of water.

Miscellaneous Events in Era 203

The Seneca Falls Convention and the Equal Voting Rights Movement

Lucretia Mott and Elizabeth Cady Stanton were two women actively involved in the campaign to end slavery. They met at the 1840 World Anti-Slavery Convention in London. As women, Mott and Stanton were barred from the London convention floor. This indignity motivated them to seek equal rights for women. They thus founded the women's rights movement in the United States, and they organized the first women's rights convention held in the States.

On July 19, 1848 two hundred women convened at the Wesleyan Chapel in Seneca Falls, New York. Elizabeth Stanton read a document she had drafted entitled "Declaration of Sentiments and Grievances." Stanton's declaration followed the ideas of the U. S. Declaration of Independence, and its preamble stated, "We hold these truths to be self-evident: that all men and women are created equal; that they are endowed by their Creator with certain inalienable rights . . . " The Declaration of Sentiments and Grievances then described injustices to women in the United States and called upon women to organize and to petition for their rights.

Men were invited to attend the second day of the convention. About forty men attended, including Frederick Douglas, the famous African American abolitionist. The Declaration of Sentiments and Grievances was adopted. The convention also passed twelve resolutions calling for equal rights for women. Eleven were passed unanimously, but one – calling for the right to vote – met opposition. Douglas sided with Stanton in supporting female enfranchisement, and the resolution passed after a lengthy debate. Some people ridiculed the convention for proclaiming women's right to vote. But the resolution marked the beginning of the women's voting rights movement in America.

Two weeks after Seneca Falls, an even larger meeting was held in Rochester, N.Y. Annual national woman's rights conventions were held, helping to fuel the women's suffrage movement. Finally, in 1920, seventy-two years after Seneca Falls, the nineteenth Amendment was adopted as part of the Progressive movement, and women had the constitutional right to vote.

Manifest Destiny

The originator of this term is believed to be John O'Sullivan, who was editor of both the Democratic Review and the New York Morning News. He was writing about the Oregon Territory, a frontier over which the United States was eager to assert dominion. The phrase became popular. Used as a verb, "to manifest" means to exhibit, to become obvious or to reveal. As an adjective, it therefore means "having been revealed." Destiny indicates "that which has been destined," and implies purpose or fate. Though the term does not explicitly state it, the term "manifest destiny" implies that the expansion of the United States from the Atlantic to the Pacific was preordained by fate or divine intention. Thus, the term justified the methods used to continue expansion, whether it was by war with Mexico or displacement of Indian tribes.

James Polk was an ardent proponent of Manifest Destiny. When he ran for President, his campaign slogan was "54, 40 or fight!" (This was a reference to a potential northern boundary of Oregon of latitude 54 degrees 40 minutes). In his inaugural address he called the claim to Oregon "clear and unquestionable."

An 1842 treaty between Great Britain and the United States had left open the question of the Oregon Territory from the Rocky Mountains to the Pacific coast. Polk wanted to resolve the Oregon issue so the United States could turn to acquiring California from Mexico. In mid-1846, Polk's administration avoided a crisis with Britain through a compromise in which Oregon would be split along the 49th parallel.

By the time the Oregon question was settled, the spirit of Manifest Destiny and territorial expansion had driven the United States into war with Mexico. The Mexican-American War ended in 1848 with the Treaty of Guadalupe Hidalgo. The treaty added 525,000 square miles of territory to the U.S., including all or parts of what is now Arizona, California, Colorado, Nevada, New Mexico, Utah and Wyoming.

The term Manifest Destiny imbued U.S. expansion with a sense of lofty idealism. But this rapid territorial expansion was accompanied by war with Mexico, as well as the by dislocation and mistreatment of Native American, Hispanic and other non-European occupants of the acquired territories.

Summary Comments about Era 203

Era 203 was an enormously significant era in many ways. James Polk's presidency is significant for his having actualized the concept of "Manifest Destiny." Via negotiation with Britain and war with Mexico, Polk established the current boundaries of the continental United States.

The Gold Rush of 1849 mobilized tens of thousands of men to travel to California to seek wealth. The Gold Rush had numerous effects. By leaving women to run businesses and manage families, the "forty-niners" unwittingly helped women experience independence. Lucretia Mott and Elizabeth Cady Stanton organized the Seneca Falls conference, initiating a movement that would eventually win rights for women.

The Gold Rush also sped up the application of California for statehood. This crystallized the hotly debated issue of whether and where slavery would be permitted and fed the flames of conflict that would bring about the Civil War.

History's emphasis on tensions leading to the Civil War may obscure progress made in medicine, technology, energy production and our understanding of the Universe. Morse's invention of the telegraph and Richard Hoe's invention of the rotary printing press vastly increased the rate at which information could be communicated. Medicine finally took a long stride forward as Louis Pasteur established the germ theory of disease. And the theory of evolution, suggested by many investigators but published by Charles Darwin, opened our eyes to a new interpretation of the history of life and the existence of our species.

The first successful use of a drilling rig to produce oil was made on August 28, 1859 by George Bissell and Edwin L. Drake near Titusville, Pennsylvania. This led to a wave of investment in oil drilling, refining and marketing. The availability of energy would lead to the world's enormous economic growth over the next one-hundred sixty years.

ERA: 204

Years: 1861 – 1885

(Civil War, First Oil Well, Robber Barons)

Presidents: Abraham Lincoln, Andrew Johnson, Ulysses S. Grant, Rutherford Hayes, James Garfield and Chester Arthur

1860				
Abraham Lincoln Hannibal Hamlin	Illinois Maine	Republican	180	40%
Stephen A. Douglas Herschel V. Johnson	Illinois Georgia	Democratic	12	29%
John C. Breckinridge Joseph Lane	Kentucky Oregon	National Democratic	72	18%
John Bell Edward Everett	Tennessee Mass	Constitutional Union	39	13%

1864				
Abraham Lincoln Andrew Johnson	Illinois Tennes-see	Union	212	55%
George B. McClelland George H. Pendleton	New Jersey Ohio	Democratic	21	45%

Abraham Lincoln

Abraham Lincoln lived from 1809 to 1865 and served as our 16th president from 1861 to 1865. He died in office when, in one of the best-known events in our history, he was assassinated by John Wilkes Booth. Lincoln's life and presidency fascinate people as much as that of any other president. This may be due in part to his ascension from backwoods obscurity to the presidency, but also to his having shepherded the country through the civil war, and to his having issued the emancipation proclamation.

Lincoln was born in a small, rustic cabin in Kentucky. When he was eight, his father moved the family to southern Indiana, where Abe helped build a log house deep in the woods. When Abe was nine, his mother died. His father remarried a year later, and he had a supportive relationship with his step-mother, who greatly encouraged his reading. He had some formal schooling, and he loved reading. At age nineteen he accompanied a store-owner on a trip down the Mississippi to New Orleans. There, he was shocked by the sight of slaves working on the docks.

At age twenty-one he helped his family move again, this time to Illinois. A year later he went on his own and moved to New Salem, Illinois, where he clerked in a store. He was said to be tall, thin, awkward and homely, but he was very strong from outdoor work. He had a reputation for kindness, consideration and honesty, and was also said to have a way of communicating with people and of inspiring trust. At age twenty-three he volunteered for the military and fought against Indians in the Black Hawk War. He was elected militia captain, a success that brought him pleasure.

At age twenty-five Lincoln was elected to the state legislature, where he took a stand against a movement to affirm that the Constitution guaranteed slavery. He studied law, won a second term in the legislature, and moved to Springfield, Illinois to practice law. In 1842 at age thirty-three, Lincoln married Mary Todd. In 1846 he won a seat in the U.S. Congress, and he identified with the Whig Party. He was disappointed with Zachary Taylor's presidency and left Congress after one term.

Lincoln empathized with impoverished persons, having been poor himself. He made speeches in favor of keeping the territories free, and came to prominence when, in 1858, he was nominated for the Senate by the Republican Party. Lincoln then engaged in a series of debates with Senator Stephen Douglas – debates that became lore. Though he lost the election to Douglas, Lincoln became famous for his performance in their debates. Due in part to that fame, the Republican Party nominated him for the presidency two years later. Lincoln won the election in the fall of 1860 but would not take office until March of 1861. During the interim, several southern states seceded from the Union, while President Buchanan did little more than hope for their return.

Lincoln took office on March 4, 1861. The political climate was ambiguous and he did not take action for a month. The tipping point came when, on April 12, 1861, southern troops attacked the Union Fort Sumter in Charleston, South Carolina. People in the northern states shed their ambivalence and Lincoln knew he could rely on their support. The following four years were among the most disruptive years in the life of the country, and the most dangerous for its survival.

Lincoln is considered to have been a great President because of what he did during the next four years. He was not without flaws and was not universally admired, but he was a sincere and effective communicator who inspired belief and confidence. The United States was the only significant democracy in the world, and Lincoln was determined to save the Union. There was a grand principle at stake: the ability of free men to govern themselves. Lincoln managed to convey this principle and its importance to the general public, and this probably enabled the country to endure a disruptive and bloody civil war.

Lincoln fought the Civil War to preserve the Union but, as the War progressed, the goal of abolishing slavery became central. In 1861 Lincoln said, "If I could preserve the Union without freeing a single slave, I would do so." Later, however, he said, "The moment came when I felt that slavery must die that the nation might live." Thus, on September 22, 1862 he issued the famous Emancipation Proclamation. Contrary to popular myth, his proclamation did not free a single slave. It would not take effect until January 1, 1863 and, even then, it applied only to states still in rebellion against the Union. Nonetheless, the Emancipation Proclamation was of great significance. Britain and France were considering entering the war on the side of the South. This would almost certainly have resulted in southern independence. This would have suited the European powers, as a divided nation would have been more vulnerable to influence and control. And, support of the South would have been an advantage relative to access to cotton. Great Britain had, however, itself ended slavery, and Lincoln's Emancipation Proclamation made support of the South less likely on ideological grounds.

Lincoln was very legalistic in his thinking. He did not think the Constitution, as written, gave the Union the right to end slavery. Thus, he worked to amend the Constitution, which was accomplished by the thirteenth amendment.

There are descriptions of Lincoln during the Civil War, walking the streets in any weather, alone and appearing lost in thought. (Note the huge difference between the 1860s and today. Then, Lincoln could walk the streets alone. Today, he would be accompanied by a secret service detail). Lincoln was troubled by the enormous cost the war inflicted in terms of loss of human life and of suffering. The outcome of the war was for a long time in doubt. By the fall of 1864, however, it had become clear that the North was prevailing and Lincoln was elected to a second term.

Lincoln's dream was to restore the Union peacefully and without revenge, and he made an impassioned plea for this at his inaugural address. He was not to live to actualize his dream. On

April 9, 1865 General Robert E. Lee surrendered to Lincoln's general, Ulysses Grant, in a small village named Appomattox Courthouse. Five days later Lincoln was shot and killed as he and his wife attended a play. Lincoln was persuasive and instilled confidence and trust. Had he lived, he may have actualized his dream of re-uniting the country peacefully and without bitterness and revenge. However, men whom historians have characterized as "among the worst" came to power in the North and the South, and the era that followed was far from peaceful.

Andrew Johnson

Andrew Johnson lived from 1808 to 1875 and served as our 17th president from 1865 to 1869. Andrew grew up in a family of modest means. His parents worked in a tavern – his father as a handyman and his mother as a maid. His father died when Andrew was three, and his mother struggled to feed the family. Johnson did not go to school and was a tailor's apprentice. He disliked his master and he re-located several times to towns in the Carolinas, but eventually settled in Greenville, Tennessee.

Johnson started a business as a tailor and did well. He married Eliza McCardle, who helped him with reading and writing. He was ambitious and made great efforts to attend a school, where he became an aggressive debater. He was loud, blunt, and unrefined. His style appealed to his community on the Tennessee frontier, where he debated public issues with the townspeople.

Johnson was a dutiful man and brought his mother and siblings to live with him. In some ways he believed in helping the poor and weak against the wealthy and strong. Johnson's political career began with a series of steps. He was elected to town council, to mayor, to Congress and to the U.S. Senate. He believed in the right to own slaves and, when he became successful, he owned several and never freed them. As a Democrat, he usually voted with the South.

After Lincoln became President, southern states were leaving the Union, and Johnson's state of Tennessee went with them. Johnson was the only southern senator to remain in the Senate. He commented that although he had spoken and voted against Lincoln, he loved his country. He called on all patriots to save the Constitution and preserve the Union. Johnson thus won favor in the North. The Republicans wanted to be seen as a party for the entire country and not just for a section of it, so they nominated Johnson for Vice President in 1864. They did not anticipate Johnson ascending to the Presidency via assassination.

A time of conflict and tumult followed Lincoln's death. The political landscape was a mélange of positive efforts to forge a better future as well as punitive and misguided efforts. Johnson agreed in principle with Lincoln's plan to reunite the country without revenge, and he made a reconstruction plan. But there were many forces lined up against that plan, and he was unwilling to work with Congress on compromises. Radical Republicans wanted the South to suffer for its secession and did not admit newly elected southerners to Congress. The South fueled the flames of conflict as many states passed several cruel laws – the "Black Codes" - severely limiting the rights of blacks.

In March 1865, Congress established the Freedman's Bureau to help former black slaves as well as impoverished white persons to re-establish themselves after the war. President Johnson did not take issue with the Black Codes, and in 1866 he vetoed an act to extend the Freedman's Bureau. Congress divided the South into military districts and placed them under military command. This atmosphere of enmity and conflict resulted in one of the more shameful episodes in the nation's history: the birth of the Ku Klux Clan. The clan intimated and killed blacks, as well as northerners who had come to the South, whether they had come south to profit or to teach. Johnson made speeches asking for moderation, but he lacked Lincoln's ability to communicate and persuade. He called for moderation, though neither his actions nor the tone of his speeches showed such a commitment.

Johnson and Congress fought about the aftermath of war and re-integrating the South. Further, Johnson removed Edward Stanton as Secretary of War, and Congress disputed his authority to do so. Angered by Johnson's removal of Stanton and other actions, Congress voted to impeach him, but the Senate fell one vote short of removing him from office. Johnson's term ended soon after that failed attempt. The nation had not heard the end of him. Six years after his return to Tennessee, he was re-elected to the Senate. Although many of his colleagues applauded him on his return, he was not gracious. He gave a strident speech defending himself and lashing out at his detractors. It was his last speech, as he became ill and died shortly thereafter.

1868				
Ulysses S. Grant Schuyler Colfax	Illinois Indiana	Republican	214	53%
Horatio Seymour Francis P. Blair	New York Missouri	Democratic	80	47%

1872				
Ulysses S. Grant Henry Wilson	Illinois Massachusetts	Republican	286	56%
Horace Greeley B. Gratz Brown	New York Missouri	Democratic / Liberal Republican	*	44%

Among the minor-party candidates were James Black, (National Prohibition), Victoria C. Woodhull (People's Equal Rights) and William S. Groesbeck (Independent Liberal Republican).

Because Greeley died before the Electoral College convened, 63 of his 66 electoral votes were distributed among other candidates.

Ulysses S. Grant

Ulysses S. Grant, born Hiram Ulysses Grant, lived from 1822 to 1885 and served as our 18th president from 1869 to 1877. Grant was born in Ohio. His father was a farmer and owned a tannery. Grant attended schools in Kentucky and Ohio, but he preferred horseback riding to school and was a poor student. Grant's father persuaded a congressman to sponsor his son to the military academy at West Point. The congressman thought his name was Ulysses Simpson Grant, and Grant was registered at West Point with that name. At West Point he was a talented horseman and set an equestrian jump record. He disliked discipline and often received demerits due to tardiness and appearance. He graduated West Point 21st in a class of 39, and later served in the Mexican War as a lieutenant under General Zachary Taylor. Grant did not approve of the war with Mexico, believing it was an artifice to acquire land. He would later write in his memoirs, "We were sent to provoke a fight, but it was essential that Mexico should commence it."

Grant's early years were not spectacular. When his commanding officer insisted that he either quit drinking or resign, he resigned. He was seen as a wastrel, drank too much, and failed at employment. By the age of thirty-seven he had failed at farming and as a rent collector for his wife's cousin. He was desperate and appealed to his father for help. He moved to Galena, Illinois where he worked as a clerk for his father's store. There, he was known by the town people as walking in a stoop-shouldered manner with a vacant facial expression.

But war would re-invigorate Grant's life. The Civil War had begun and Grant began enlisting and drilling men as volunteers. He reached out to some officers he had known and obtained help from a congressman. The North needed trained officers, and he was appointed as colonel and later promoted to general. President Lincoln was having difficulty settling on a commander for all the Union forces, and Grant was given that command in 1864.

Grant succeeded as a General, though not everyone approved of his methods. He had more men and equipment than did the Confederates, and he was willing to suffer many casualties to achieve victories. Nonetheless his victories led to the end of the Civil War, and he thus became a national hero. Successful military leaders often capture the public's imagination, and in 1868 the Republican Party made him their nominee for President. Grant had no interest in politics, and it is said that he had only once even cast a vote in a presidential election. Grant's opponent in the 1868 election was New York Democratic governor Horatio Seymour. Grant himself remained largely silent during the campaign. His supporters campaigned against the Democrats as having been "the party of secession," and Grant was elected.

At his inauguration, Grant called for government to protect life and property for all. The fifteenth amendment had been passed by Congress prior to Grant's inauguration. It prohibited the federal government and the states from denying citizens their right to vote based on "race, color or previous condition of servitude." Essentially, the amendment gave African American men the right to vote. Grant urged the states to ratify the amendment, and it was ratified on February 3, 1870.

Despite a promising start, Grant's presidency is usually considered by historians to have been a disaster. It was not due to a lack of desire to be a good president, but rather due to a lack of ability and to poor judgment in appointing officials. His administration was riddled with corruption, and his appointees used their positions to line their own pockets. Though not gifted, Grant was honest. But he was naïve and could not believe that the corruption was going on, even when he was apprised of it.

The reconstruction of the South was the major problem facing the nation. Despite the desire of President Lincoln to navigate this process without revenge, radical Republicans in Congress passed a series of very harsh laws, and sent troops to southern states to enforce the laws. The South retaliated with equally harsh and restrictive laws against blacks.

Grant did manage a few accomplishments of note. Despite what historians regard as many disastrous appointments, Grant's appointment of Hamilton Fish as Secretary of State is regarded as having been a good one. Cuban rebels were fighting Spain, and many Americans wanted to support the rebels. Grant and Fish believed this would be unwise, partially because the nation had been weakened by the Civil War. They felt that avoiding a possible war with Spain was important. Grant made his position clear to Congress and a resolution to support the rebels was defeated.

Secretary of State Fish, on President Grant's behalf, resolved some major issues with Great Britain. English warships had disrupted Northern shipping during the Civil War, and this was in violation of its supposed neutrality. The United States claimed Britain owed compensation, known as the Alabama Claims. In addition, the United States and Britain disagreed over issues regarding Canada, including boundary disputes and fishing rights. A Joint High Commission of American, Canadian and British negotiators met in Washington, D.C. to forge an agreement, and they succeeded. Most of the issues were resolved, and the Alabama Claims were submitted to international arbitration. The result was the Treaty of Washington, which included a provision that Britain owed the United States 15.5 million dollars. The Senate approved the treaty, which legitimized the viability of international arbitration and improved relations between Britain and the United States. This treaty may have been the best accomplishment of Grant's presidency.

Grant had some admirable traits. He was said to have been gracious to those whom he defeated in the Civil War. Further, due to his support of the fifteenth amendment, some observers have referred to him as "the first civil rights President." He owned a slave, but freed him before he had obtained financial success. Grant retained his popularity when his second term ended. Though people recognized the corruption of his administration, he was not seen as responsible for it, and he was well received during a world tour. He needed money and invested everything in a banking business. He left management of the business to "friends," and these friends were like those to whom he had entrusted government posts and who had been corrupt. The business went bankrupt and Grant was in debt. He had always been a heavy smoker and developed throat cancer. Mark Twain offered money to Grant if Grant would write his autobiography. Grant was determined to fight pain and his impending death and to finish the work. He did so a week before his death and his memoirs earned a fortune for his family.

1876				
Rutherford B. Hayes William A. Wheeler	Ohio New York	Republican	185	48%
Samuel J. Tilden Thomas A. Hendricks	New York Indiana	Democratic	18	51%

Rutherford B. Hayes

Rutherford Hayes lived from 1822 to 1893 and served as our 19th president from 1877 to 1881. Rutherford grew up in Ohio. His father had died prior to Rutherford's birth, and his elder brother Lorenzo later died while skating on an icy pond. Rutherford loved fishing, hunting, rowing and swimming, but he was also a serious student. He had an affinity for public affairs from an early age. He particularly admired George Washington and he memorized patriotic speeches.

College education had been a rarity for U.S. Presidents, but Hayes graduated from Kenyon College and then went to Harvard Law School. As his graduation from Harvard Law School approached, he wrote in his diary, "The rudeness of a student must be laid off, and the quiet, manly deportment of a gentleman put on." He returned to Ohio and opened a law practice in Cincinnati. His practice was slow at the start, but he won some major cases and eventually had a thriving practice. In 1858 Hayes was appointed city solicitor of Cincinnati, and in 1859 he was elected to a two-year term.

As a youth, Hayes had dreamed of being a military leader and, though he had no military training, he was appointed captain of volunteers when the Civil War began. He became a courageous and successful officer, and is said to have been wounded four times and to have had four horses killed under him. He had advanced to major general by war's end.

In 1864 the Republican Party recruited him as a candidate for Congress. Though he refused to leave his post to campaign, he was nonetheless elected. He was elected governor of Ohio after serving two terms in Congress. As governor, he backed safety codes for mines, a state civil service system, voting rights for blacks and improvements in prisons and mental hospitals. He asked the federal government to limit issuing paper money in order to avoid inflation. He founded Ohio State University and worked to reform state government. He was nicknamed "Old

Granny" by members of his party who were inconvenienced by his attempts to make government more honest and accountable.

As the 1876 election year approached, the Grant administration was widely recognized as corrupt. Some members of a divided Republican Party thought that their only path to victory was to nominate a candidate with a solid reputation for honesty, and that person was Rutherford B. Hayes. The election took a very unusual turn. Democratic candidate Samuel Tilden won the popular vote 4,284,020 to 4,036,572. And, Tilden had 184 votes in the electoral college to Hayes' 165. However, 185 electoral votes were needed to win, and Tilden was one short. The votes of four states were in dispute. If Tilden were to get a single vote from any of the disputed states, he would win the presidency. The argument went on for months in Congress. Finally, a deal was struck. The Democrats agreed that a committee would decide the presidency, and the Republicans agreed that federal troops would be removed from the South. The committee voted along party lines – eight to seven – and Hayes became President.

Hayes wasted no time signaling his intent to push for a civil service in which appointments would be made based on ability rather than on partisanship. In his inaugural address he urged "thorough, radical and complete" civil service reform. Hayes kept his party's promise and federal troops were withdrawn from South. He required assurances from the governor of South Carolina that the constitutional rights of African Americans would be upheld. Hayes acted similarly in Louisiana. He received promises from that state's governor that friendship between the races would be encouraged via constitutional amendments. The long, contentious, disruptive period peculiarly named the "Reconstruction Era" was, at least to a degree, over.

President Hayes won a long and contentious battle over the New York Customhouse. Seventy-five percent of the county's tariffs were collected there, and money was routinely funneled to the political machine of those in power. In 1878 Hayes suspended Chester Arthur and his associate, Alonzo Cornell, who essentially worked for Senator Roscoe Conkling's political machine. Hayes later wrote, "The great success of my administration was in getting control of the New York Customhouse and in changing it from a political machine for the benefit of party leaders into a business office for the benefit of the public."

President Hayes put effort into presiding over an honest administration. He appointed honest men and removed men who used their positions for personal gain. Hayes wanted to eliminate the corrupt spoils system. Although the Hatch Act of 1939 was almost seventy years away, Hayes issued an executive disorder making it illegal to require workers to contribute money to candidates. His order forbade civil servants from participating "in the management of political organizations, caucuses, conventions or election campaigns." Congress did not go along with all of Hayes' attempted reforms. But he strove mightily to have a more honest government and historians indicate that he left government at least modestly more honest than he found it.

Rutherford Hayes and his wife Lucy had no regrets about his having made a pledge to be a one-term President. In planning his retirement, Hayes decided that he would "promote the welfare and the happiness of his family, his town, his State, and his country." Hayes was an egalitarian and continued his support of social causes he had already championed.

Hayes believed that the American government could be no better than its people. He contended that the country's ethical and material life depended on an educated public. Thus his major thrust was to see that public education received universal tax support. He championed improved opportunities for students at all educational levels. He fought for federal subsidies for children of all races in poor school districts. He believed that education had the potential to improve the economic status of the poor and to enlighten the intolerant.

Hayes also had an interest in prison reform and reducing crime, and he opposed the death penalty. He believed that crime was the product of poverty and desperation, and that criminals could be reformed through education.

Hayes believed many of society's problems could be reduced by a fairer distribution of wealth. He saw the gap between "rich industrial kings" and laborers as the greatest problem facing the nation. Thus, he favored federal regulation of industry. He believed in industrial education so that the rich would know what it meant to toil, and he supported inheritance taxes to equalize wealth. He recognized social Darwinism for what it was. He believed unregulated competition resulted not in the survival of the fittest but in the triumph of the most predatory corporations. He worried deeply about "a government of corporations, by corporations, and for corporations." Hayes used his retirement years for productive work in support of humanitarian aims. Hayes died of heart disease on January 17, 1893.

1880				
James Garfield Chester A. Arthur	Ohio New York	Republican	214	48.3%
Winfield S. Hancock William H. English	Pennsylvania Indiana	Democratic	155	48.2%

James A. Garfield

James Garfield lived from 1831 to 1881 and served as our 20th president from March until September of 1881. He was born into a poor family on a frontier farm in Ohio, and his father died when James was two years of age. His mother was described as "bright and cheerful" and as trying to maintain a happy home atmosphere. But the family could not escape from poverty and much of young James' efforts went into helping on the family farm.

Garfield possessed unusually strong intellectual ability, but he was a robust young man who loved the outdoors. In his early years he worked with his hands, first on his mother's farm and then on canal boats. The story is he fell overboard fourteen times, contracted a severe fever, and returned home resolving to use his brains instead of his brawn.

Though indifferent to religion in his earlier years, he joined a church at age nineteen, and from that point on always saw divine will at work. Garfield worked carpentry jobs that allowed him time to attend school, and he was a prodigious learner. He attended Western Reserve Eclectic Institute (later Hiram College), where he was well-liked, a serious student, and developed his ability in public speaking. At age 23 he enrolled at Williams College. He graduated from Williams and returned to Western Reserve as a professor of classical languages. His intellect was so sharp that he is said to have been able to simultaneously write Latin with one hand and Greek with the other. He became disenchanted with higher education due to bickering among faculty, and he sought a new career path. At age twenty-eight he ran as a Republican for a seat on the state senate, and he won.

Garfield was an ardent abolitionist who opposed slavery in the territories, and he campaigned for Abraham Lincoln in the 1860 presidential election. In 1861 he organized the 42nd Ohio Infantry and quickly rose to the rank of full colonel. He had no military background, but read everything he could on military strategy, including accounts of European military leaders such as Napoleon. He became an excellent officer and won distinction twice during the Civil War. In January 1862 he defeated a larger force in the Battle of Middle Creek, and in 1863 he made a daring ride under enemy fire at Chickamauga. He was elected to Congress from Ohio, but was reluctant to leave the army.

President Lincoln convinced him to take his seat in Congress, where he served for eighteen years.

During his years in Congress Garfield learned to temper his earlier, more strident views, and developed the ability to work for compromise. In the post-Civil War years, he opposed his more radical colleagues' desire to punish the South, and he favored Lincoln's call for moderation.

The next years were a vital time in the life of U.S. democracy. Government was not generally seen as serving the people, but rather was seen as a vehicle for politicians to make themselves and their friends wealthy. This was true under Grant's administration, reportedly due to naiveté and not dishonesty on Grant's part. President Hayes' efforts had yielded only modest improvement. What is sometimes known as the "patronage system" and sometimes as the "spoils system" was rampant. There were hardly any regular civil service employees, and almost all government jobs were handed out to the victorious party's supporters. This was still in effect when Garfield was elected President in 1880.

Ironically, though it was Garfield's intent to reform the system and to restore honesty, as well as the public's faith in government, he was shot and killed by Charles Guiteau, a man to whom he had denied a job. Guiteau fired two shots. One grazed Garfield's arm and one lodged near his spinal column. Medical techniques were crude and doctors were unable to locate the bullet. It is likely that instruments used to probe for the bullet were contaminated, and Garfield may have lived if the wound had been left alone. He was brought to a New Jersey seaside village for fresh air, where he died on 19 September 1881, having said, "My work is done."

Garfield's death shocked the nation, and the public began to demand a better civil service system that would keep honest workers at their posts. Congress grudgingly acceded to the public's demands and civil service laws were improved. This is

Garfield's legacy, and it may be underestimated in terms of its importance to the continuation of the nation's experiment in democracy.

Chester A. Arthur

Chester Arthur lived from 1829 to 1886 and served as our 21st president from 1881 to 1885. His father was a Baptist clergyman and ardent abolitionist who moved the family to various towns in Vermont and upstate New York. Chester attended Union College in Schenectady in 1845, graduated in 1848, and then studied law and taught school. Chester re-located to New York City where he clerked in a law firm, continued his study of law, and was admitted to the bar in 1854. His law career has been called "unspectacular," but he won a noteworthy civil rights case. He represented an African American woman named Elizabeth Jennings, who had been assaulted when she would not vacate a streetcar reserved for whites. A law had been enacted that forbade denying "colored persons" access to public transportation. Arthur argued that her treatment had violated that law and he won a cash settlement for his client. More importantly, New York began integrating its streetcars to comply with the new law.

Arthur's career and presidency may be seen as an example of the phrase "the office makes the man," and of how a person's values and beliefs can sometimes change. Politics in the mid-to-late 1800's were openly partisan, opportunistic and corrupt. Important jobs were handed out to men who supported the party in power, who in turn contributed some of their pay back to the party. This system is known as the "patronage system," or the "spoils system." It was accepted as a common practice, and Chester Arthur was an avowed supporter of it. President Grant appointed Arthur to the very important political position of Collector of the Port in New York. Working for Senator Roscoe Conkling, Arthur used the position to give jobs to people who would work for and vote for the Republican Party. This system was fervently supported by Republicans known as "stalwarts."

However, there was another faction in the Republican Party – a group who did want reform and more honest government. Chester Arthur was a stalwart and supported a third term for Grant. When James Garfield was nominated for the presidency, Arthur was nominated for Vice President as a compromise.

President Garfield had favored significant reform for the civil Service. He believed in written examinations for government jobs as part of a system in which people would be hired based on their qualifications and not through the patronage system. Chester Arthur, the stalwart, openly opposed Garfield's proposals. But events were to change his views. President Garfield was murdered by a man named Charles Guiteau who, prior to shooting Garfield, is said to have written," His death is a political necessity. I am a lawyer, theologian and politician. I am a stalwart of the stalwarts."

Many stalwart politicians were glad Arthur was to be President, as they believed things would return to the ways of the Grant administration. But Arthur could not forget the reason for the murder of his predecessor. And, as President, he felt a duty not to his own party, but to all citizens. He attempted to carry on the reforms proposed by Garfield, and he asked Congress for a new civil service law. The old guard stalwart politicians were angry at Arthur's change in stance. But many Americans were demanding reform, and many of the old guard stalwarts would be defeated in the elections of 1884. In the meantime, a new civil service law was passed. In addition, Arthur worked diligently to streamline the postal system and to modernize the navy. The party stalwarts did not forgive Arthur and he was denied the party's nomination for a second term. He went home to New York with the admiration of much of the nation. He is an example of an elected official rising to the challenge of his office and altering his stance to meet the needs of the people.

Chester Arthur was diagnosed with a kidney disease called Bright's Disease in 1882, but he kept it secret. He knew his condition was fatal, and he did not ask to be nominated for a second term. Efforts were made to nominate him, but his former secretary of state, James G. Blaine won the nomination on the fourth ballot.

Due to failing health, Arthur was unsuccessful in his attempt to return to the practice of law. He became too frail even to pursue his favorite pastime of fishing. Chester Arthur died at home with his family on November 18, 1886. His successor, President Grover Cleveland, attended his funeral in Albany, New York.

Events and Themes in Era 204

Industrialization and Immigration

The Industrial Revolution was well underway, the United States was viewed as a land of opportunity, and the country experienced a surge of immigration. More than eleven million people flooded into the country from Europe and China, seeking employment and a better life for their families. The United States began the process of transformation from rural/agrarian to urban/industrial. Unfortunately, factory owners took advantage of the large supply of willing workers and lowered wages. An era of labor unions, strikes and repressive government intervention loomed on the horizon.

The Plight of Labor

This was an era in which business and industry held all the cards vis a vis labor. Wages were suppressed and working conditions were poor. Labor unions were not legal at the time, and government often used repressive means to break labor strikes. It may be said that the country was built on cheap labor, but millions of laborers and their families lived with less comfort and privilege due to suppressed wages.

The prevailing economic theories of the day stated that wages would always migrate to subsistence levels and nothing could change that. Thus, industrialists felt they had theoretical backing for their exploitation of labor.

The country's economy was subject to severe downturns occurring approximately every twenty years or less. There was an economic depression in 1876, and in 1877 the Baltimore and Ohio Railroad (B&O RR for you Monopoly players) cut wages three times in a year. Workers were not represented by trade unions. It was common for city and state governments to organize armed militias, aided by national guard, federal troops and private militias. They fought and suppressed the workers.

The Great Railroad Strike of 1877 began on July 14 in Martinsburg, West Virginia. It followed the depression of 1876, and is sometimes referred to as "the Great Upheaval." Disruption was widespread and, at its height, the strikes were supported by about 100,000 workers. Because of economic problems and pressure on wages by the railroads, workers in numerous other cities also went out on strike. In some cities, workers burned and destroyed physical facilities, engines and railroad cars.

With the intervention of federal troops in several locations, most strikes were suppressed by early August. By the end of the upheaval, an estimated 100 people had been killed across the country. Labor continued to work to unionize and to work for better wages and conditions. Fearing social disruption, many cities built armories to support their militias. These defensive buildings still stand today, not as symbols of readiness to defend our shores, but as symbols of the effort to suppress the labor unrest of this period.

"Robber Barons" or "Captains of Industry"?

Professor Emeritus Mark Stoler of the University of Vermont describes a game he would play with students of American history. He asked them to name all the Presidents they could remember prior to 1865, and they were able to name at least some. He then asked them to name all the Presidents they could remember after 1900, and they named quite a few. Finally, he asked his students to name the Presidents between 1865 and 1900, and he reports getting blank stares.

The Presidents between 1865 and 1900 are Andrew Johnson, Ulysses Grant, Rutherford Hayes, James Garfield, Chester Arthur, Grover Cleveland, Benjamin Harrison, Grover Cleveland (again), and William McKinley. Professor Stoler then suggests that there is little reason for his students to remember those Presidents. That was the era of industrialization, and the major figures of industrialization are more memorable than were the Presidents of that era.

The mid to late nineteenth and early twentieth centuries is sometimes called "the gilded age." This period witnessed the rise to enormous wealth and power of several entrepreneurs in the rapidly industrializing United States. A very few of the many well-known persons in this category were Andrew W. Mellon (finance, oil), Cornelius Vanderbilt (water transport, railroads), James Buchanan Duke (tobacco, electric power), John Jacob Astor (real estate), William Randolph Hearst (media), Jay Gould (railroads), Joseph Seligman (banking), John D. Rockefeller (oil), Henry Ford (automobiles) and Andrew Carnegie (steel).

The motives of these industrialists are clear: wealth and dominance. There is debate, however, on how history should portray them. Some refer to the powerful industrialists of the gilded age as "robber barons," portraying them as ruthless businessmen who would use any means to achieve great wealth. These "robber barons" were accused of exploiting the working class and forcing terrible working conditions and unfair labor practices on workers.

Another view of the industrialist is that of "captain of industry." This term views these men as daring and energetic leaders who transformed the American economy with their ingenuity and vision. Many, though certainly not all, of these industrialists turned their attention to philanthropy later in their lives. John D.

Rockefeller, Andrew Mellon and Andrew Carnegie are three notable examples. When they did turn to philanthropy, what were their motives? Was it love of humanity, or was it concern about their own reputations and legacies? Or, could they have been more ruthless and exploitative in their younger years, and then developed philanthropy as a result of genuine self-actualization after their needs for dominance and wealth had been met?

The debate over robber baron versus captain of industry may be a debate over the values of our nation. Industrialization has brought both advantages and disadvantages to humankind. No attempt is made here to judge the ultimate value of industrialization, as that is left to readers to explore.

John D. Rockefeller and Standard Oil

The story of the rise to dominance of Standard Oil, and a few other major American industrial corporations, represents an extremely significant series of events in the development of American industrial strength, America's rise to international economic dominance and, ultimately, in America's preservation of competition through antitrust legislation.

Standard Oil was stablished in Ohio in 1870 by John D. Rockefeller and Henry Flagler. it was the world's largest oil refinery at the time. A grouping of companies that function cooperatively under one corporate structure is known as a business trust. In 1882, Rockefeller consolidated several companies to create the Standard Oil Trust, thereby becoming an innovator in the development of the business trust. Rockefeller controlled all phases of production, refining, transportation, distribution, and marketing of the industry. The trust streamlined production, lowered costs, and allowed it to undercut competitors. Standard eventually controlled almost ninety percent of the country's oil production. It was the largest oil refiner in the world and one of the world's first multinational corporations.

Standard's secret deals with railroads lowered kerosene prices from 58 cents in 1865 to 26 cents 1870. Consumers liked lower prices but competitors suffered from the company's practices. Critics accused Standard Oil of using aggressive pricing to monopolize the industry and destroy competitors.

Standard Oil's controversial history as one of the world's first and largest multinational corporations ended in 1911, when the United States Supreme Court ruled that Standard Oil was an illegal monopoly. Standard Oil trust was re-organized into thirty-four smaller, independent companies. As the income of these individual enterprises proved to be greater than that of a single larger company, Rockefeller became the richest man in the world. By 1896, Rockefeller had disengaged from business to concentrate on his philanthropy.

Henry Ford and the Ford Motor Company

When Henry Ford was 13 years old, his father gave him a pocket watch. Young Henry promptly took it apart and reassembled it. Neighbors were impressed and requested that he fix their timepieces. Ford left home at age 16 to be an apprentice machinist at a shipbuilding firm in Detroit. Subsequently, he studied bookkeeping and learned to operate and service steam engines. Henry Ford was hired as an engineer for the Detroit Edison Company in 1890. In 1893, his talents earned him promotion to chief engineer.

All the while, Ford worked on plans for a horseless carriage. In 1892, he built his first gasoline-powered buggy, with a two-cylinder, four-horsepower engine. In 1896, he presented his plans to Thomas Edison. Edison encouraged Ford to continue to improve his invention. In 1899 he raised money from investors, left Edison, and pursued his dream of a car-making business. After a few trials building cars and companies, Henry Ford established the Ford Motor Company in 1903.

In 1908, Henry Ford introduced the Model T. Earlier models had been produced at a rate of only a few a day, with groups of two or three men working on each car from components made to order by other companies.

Henry Ford was an innovator and introduced the world's first moving assembly line. This reduced assembly time for a chassis from 12.5 hours to 2 hours, 40 minutes, and ultimately to 1.5 hours. This boosted yearly output to over two hundred thousand cars. Ford promised his workers profit-sharing if sales hit 300,000 between August 1914 and August 1915, and that figure was reached. By 1920, production would exceed one million per year.

His innovations increased productivity but were hard on workers. In January 1914, Ford solved his employee turnover problem. He established an 8-hour day and a 5-day work week, which became a standard in American industry. He doubled his workers' pay to five dollars per day (reducing his popularity among other industrialists). Ford reasoned that if he increased wages, his workers could also be the company's customers. Sales increased, as a line worker could buy a Model T with less than four months' pay. Ford also instituted hiring practices that identified the best workers, including disabled people considered unemployable by other firms.

Henry Ford was a great innovator, but history also knows him as a virulent anti-Semite who blamed the Jewish people for things for which they could not possibly have been responsible. He is one of many examples in our history of a person or movement with a combination of positive and negative characteristics.

<u>Andrew Carnegie and Carnegie Steel (later U.S. Steel)</u>

Andrew Carnegie was a Scottish-American industrialist, business magnate and, later, a philanthropist. Andrew was born in 1935 in Dunfermline, Scotland, the son of a weaver and a shoemaker's sewer. His family desired opportunity and, when Andrew was thirteen, they moved to America. Andrew began his career as a laborer in a cotton factory.

Andrew was an ambitious young man who worked at a series of jobs, each a better position. At age twenty-four he attained the position of superintendent for Pennsylvania Railroad's Pittsburgh division. In this position, he earned sufficient money to invest in such industries as coal, iron and oil, as well as railroads. Railroads were a growth industry, and Carnegie rode the wave of growth by founding ventures such as an iron bridge building company and a telegraph firm. Carnegie became a very wealthy man by his early thirties. In the early 1870s, Carnegie co-founded his first steel company and, over the next few decades, created a steel empire.

Carnegie led the expansion of the American steel industry and made his fortune through aggressive investment and innovations. He increased production efficiency through a controlled and rapid burning away of the high carbon content of pig iron during steel production. This process, originally conceived by Sir Henry Bessemer, had been further developed by other innovators. His second great innovation was vertical integration of the industry. Carnegie minimized inefficiencies by investing in raw materials, transportation and production. In 1892 Carnegie combined his assets and those of his associates and launched the Carnegie Steel Company. By the late 1880s, Carnegie Steel was the largest manufacturer of pig iron, steel rails, and coke in the world.

Carnegie considered himself a champion of the working man, but there are stains on his record in this regard. In 1892, at his Homestead, Pennsylvania steel mill, workers went on strike to protest wage cuts. Carnegie, on vacation in Scotland during the strike, put his support in his general manager Henry Clay Frick, who called in 300 Pinkerton armed guards to protect the plant. A bloody battle broke out between the workers and the Pinkertons, leaving at least 10 men dead. The state militia was brought in to take control of the town, union leaders were arrested, and Frick hired replacement workers for the plant. After five months, the strike ended with the union's defeat. The labor movement at Pittsburgh-area steel mills did not recover for four decades. In addition, though reports of actual fatality rates differ, there was a considerable fatality rate for workers in steel mills.

In 1901, J. P. Morgan purchased Carnegie Steel for $480 million, making Carnegie one of the world's richest men. During the last two decades of his life, Carnegie gave away almost 90 percent of his fortune to charities, foundations, and universities. His 1889 article proclaiming "The Gospel of Wealth" advanced the practice of philanthropy and called on the rich to improve society. He is quoted as having said, "The man who dies rich dies disgraced."

Inventions in Era 204

The light bulb

The importance of electric lighting cannot be overstated. Prior to it, for instance, factories could only operate by daylight. Thomas Edison was born in Ohio in 1847 and he died in 1931. He was a prolific inventor who had over a thousand patents by the end of his career. In 1878, Edison focused on inventing a safe, inexpensive electric light to replace the gaslight – a challenge that scientists had been grappling with for 50 years.

With the help of prominent financial backers, Edison set up the Edison Electric Light Company and began research. The invention of the lightbulb was no simple task. At his New Jersey laboratory, he and his associates tried three thousand ideas to develop a usable incandescent light. His idea was to pass electricity through a filament to make it glow, and to house the filament in a bulb with partial vacuum to slow the burning of the filament. Edison said, "Before I got through, I tested no fewer than 6,000 vegetable growths, and ransacked the world for the most suitable filament material." In the summer of 1880, he hit on carbonized bamboo as a viable material for the filament, which proved to be the key to a long-lasting and affordable light bulb.

In 1881, he set up an electric light company in Newark. Electric lighting was one of the major factors that changed the world as the United States would transition its primary economy and lifestyle from rural and agrarian to urban and industrial.

Medical Advances in Era 204

Antiseptics

In today's world we grow up being frequently told to wash our hands and to be aware of "germs." It is interesting to note that it was only near the end of the term of our eleventh President that the utility of antiseptics was recognized. And the germ theory of disease was not developed until after the terms of twenty-one U.S. Presidents.

Ignaz Philipp Semmelweis, a Hungarian physician, lived from 1818 to 1865. He is known as a pioneer of antiseptic procedures. Semmelweis worked in obstetrical clinics, where there was a high incidence of puerperal fever, also known as "childbed fever." Semmelweis discovered that the incidence of this disease could be drastically cut using hand disinfection. As obvious as this is to us with our current-day "germ awareness," it was new and not well-accepted then. At Vienna General Hospital's First Obstetrical Clinic, doctors' wards had three times the mortality of midwives' wards, where the midwives did practice hand-washing. In 1947, Semmelweis proposed the practice of washing hands with chlorinated lime solutions. Some observers dubbed him the "savior of mothers." Semmelweis published his findings in his book *Etiology, Concept and Prophylaxis of Childbed Fever*.

New ideas are frequently met with resistance and scorn by promoters of accepted practices. This was true with Semmelweis' discovery. Despite reports of hand washing reducing mortality to below 1%, Semmelweis's observations clashed with established medical opinions of the time. His ideas were rejected by the medical community. Semmelweis could offer no acceptable scientific explanation for his findings, and some doctors were offended by the idea that they should wash their hands. It was not until Louis Pasteur confirmed the germ theory of disease fifteen years later that Semmelweis's ideas earned widespread acceptance. Semmelweis' misfortune was not limited to being denied credit for his revolutionary discovery. In 1865, at age 47, he contracted blood poisoning, was committed to an asylum, and died after being beaten by the guards.

What We Learned about the Universe in Era 204

The Beginnings of the Science of Genetics

Gregor Mendel, an Austrian monk and scientist, is generally recognized as the founder of the science of genetics. Farmers and animal breeders had known for a long time that you could enhance desired traits through careful crossbreeding. But this knowledge had never been systematized prior to Mendel. Mendel conducted experiments with pea plants. From 1856 to 1863 he formulated many rules of heredity, now referred to as the Mendelian laws of inheritance.

Mendel worked with seven characteristics of pea plants: plant height, pod shape and color, seed shape and color, and flower position and color. Taking plant height as an example, Mendel showed that when a tall plant and a short plant were cross-bred, their offspring always produced tall plants. The trait of shortness seemed to have disappeared. But when he bred the resulting tall plants, there was a surprise. In the next generation, short plants reappeared in one of four plants. Mendel realized that the trait of shortness had not disappeared. Rather it had been suppressed. To explain this, Mendel coined the terms "recessive" and "dominant" to refer to certain traits. In the preceding example, the short trait, which seems to have vanished in the first filial generation, is recessive and the tall is dominant. He published his work in 1866, indicating that certain invisible factors were at work. We now call these factors genes. Mendel had given us a new understanding of inherited traits.

The profound significance of Mendel's work was not recognized until the turn of the 20th century, more than thirty years later. Several researchers verified Mendel's findings, ushering in the modern age of genetics.

The Restless Earth in Era 204

<u>1871 The St. Louis Tornado</u>

At 3:00 p.m. on March 8, 1871 an F-3 tornado touched down in St. Louis, Missouri. It was extremely fast, traveling east-northeast at 70 miles per hour. Though only on the ground for 3 minutes, it cut a swath 250 yards wide and 5 miles long into the landscape. Thirty homes were destroyed and 30 more were severely damaged. Nine people were killed and 60 were injured. It was the first of four tornadoes that hit the business district of St. Louis from 1871 through 1959.

<u>1883 Krakatoa</u>

The most notable eruptions of the Indonesian volcano Krakatoa occurred August 26–27, 1883 and were among the most violent volcanic events in recorded history. The eruption was equivalent to 200 megatons of TNT — about 13,000 times the power of the Little Boy bomb that devastated Hiroshima, Japan, in World War II, and four times the power of the most energetic nuclear device ever detonated.

The 1883 eruption ejected approximately 6 cubic miles of rock and the explosion was heard 3,000 miles away. According to the records of the Dutch East Indies colony, 165 villages and towns were destroyed near Krakatoa. At least 36,000 people died, and many thousands were injured, mostly from the tsunamis that followed the explosion. The eruption destroyed two-thirds of the island of Krakatoa.

Eruptions in the area since 1927 have built a new island, Anak Krakatau (Indonesian for "Child of Krakatoa"). Periodic events have continued, with recent eruptions in 2009, 2010, 2011, and 2012. In late 2011, this island had a radius of roughly 1.2 mi, and a high point of about 1,063 ft above sea level, growing 16 ft each year. In 2017 the height of Anak Krakatau was reported to have reached 1200 feet above sea level. The process by which the Earth releases its primordial heat via volcanic and seismic activity continues unabated.

Miscellaneous Events in Era 204

Intercontinental Communication

Although President Lincoln and Queen Victoria had exchanged telegraph messages via transatlantic cable in 1858, that cable parted after two months. In 1867 the transatlantic telegraph cable was successfully re-connected.

Dow Jones

Dow Jones & Company was established in 1882 by Charles Dow, Edward Jones and Charles Bergstresser. Their first product was a daily series of brief news bulletins, hand-delivered throughout the day to traders at the stock exchange. Over the course of two decades, they created three products that have been emblematic of financial journalism: The Wall Street Journal, Dow Jones Newswires and the Dow Jones Industrial Average.

The Wall Street Journal was first published July 8, 1889. The afternoon newspaper covered four pages and sold for two cents. In the first issue the publishers expressed a commitment to excellence: "We appreciate the confidence reposed in our work. We mean to make it better." The Dow Jones Industrial Average was officially launched in 1896. It is still the most frequently reported stock market index, though other averages such as the Standard and Poor's 500 are considered more representative of the entire market.

Summary Comments on Era 204

The years of Era 204, 1861 through 1885, were very eventful. The Civil War was fought, with immense numbers of casualties. Before his assassination, President Lincoln made an impassioned plea to re-unite the country without recrimination. Although his successor, Andrew Johnson, had the same aim, he lacked Lincoln's ability to influence people. The so-called "Reconstruction Era" was an era of brutal conflict that did little to heal the country's wounds. The Ku Klux Clan rose during that time, a testament to the atmosphere of enmity and hatred that persisted.

Oil had been discovered at the end of Era 203 (Seneca Falls Convention, Manifest Destiny, Trains and Canals). Oil production, and the great industrial growth it enabled, was a dominant factor in Era 204. Powerful industrialists such as John D. Rockefeller, Andrew Carnegie, Henry Ford and others were the dominant figures in the era, in many ways eclipsing political leaders. The working class did poorly in general during the era, as there was plenty of labor available, partly due to immigration. Most industries exploited labor brutally, perhaps justifying their policies with prevailing economic theories that predicted that wages would always migrate to subsistence levels. Labor unions were not legal during this era, and states and the federal government did not hesitate to use armed force to put down labor strikes.

The government under President Grant was extremely corrupt, due to Grant's unwillingness to recognize what his appointees were doing. The spoils system was rampant and widely accepted, and the public had little if any faith in government.

Congressional action was needed to decide on the election of Rutherford Hayes. The deal that was made resulted in Hayes being President and in federal troops being removed from the south, ending the so-called Era of Reconstruction. Hayes was dedicated to an improved society, and attempted with modest success to make the government more honest. Hayes' successor, James Garfield, was a man of prodigious intellect who wanted to continue President Hayes' efforts to make government more honest and to have a civil service system in which jobs would be awarded based on merit rather than on political support. Garfield was shot and killed by a man proclaiming himself to be a "party stalwart." Garfield's successor, Chester Arthur, had opposed Garfield and was known as a party stalwart who went along willingly with the spoils system, allowing elected officials to profit from their positions. But he was affected by his awareness of why President Garfield had died and by his sense that his duty was to all the people. He thus rose to the occasion, broke with the stalwarts, and tried to further the aim of more honest government.

Era 204 saw a major advance in that medical practice finally became aware of the benefits of sanitation and antiseptics. Gregor Mendel's experiments with pea plants earned him the title of father of genetics and set the stage for furthering human knowledge in that area. And, after an immense research effort, Thomas Edison hit the formula to create a useable incandescent light bulb. The August, 1883 eruption of Krakatoa in Indonesia was heard 3,000 miles away and resulted in approximately 36,000 deaths. By 1870 the U.S. population was just under forty million, and one in eight children was working.

ERA: 205

Years: 1885 – 1909

(Telegraph, Populist Movement, Pure Food and Drug Act)

Presidents: Grover Cleveland, Benjamin Harrison, Cleveland (again), William McKinley and Theodore Roosevelt

1884				
Grover Cleveland Thomas A. Hendricks	New York Indiana	Democratic	219	49%
James G. Blaine John A. Logan	Maine Illinois	Republican	182	48%

Grover Cleveland

Grover Cleveland lived from 1837 to 1908. He has the distinction of having been both our 22nd president from 1885 to 1889 and our 24th president from 1893 to 1897. He grew up in a family with many children but little money. He wanted to attend Hamilton College but his father, a Presbyterian preacher, was having financial travails and obliged Grover to quit school and go to work in a store. At age seventeen Grover wanted a better job and decided to move to Cleveland because he liked the name. But when he stopped in Buffalo to visit a relative, he found a job in a lawyer's office. He had time to study and he became a lawyer at age 22.

Cleveland's father had died prior to the Civil War and his mother was poor. Cleveland and his two brothers knew that one of the three would have to stay home to support their mother. They drew lots and Cleveland drew the short straw and stayed home to work. Cleveland was a success as a lawyer and was recruited by the Democratic Party to run for mayor of Buffalo. He won and gave the city an honest government, firing employees who accepted bribes. His reputation for saving taxpayer money and for cleaning up corruption was so strong that he won the governorship without even making a campaign speech. He continued to make sure tax dollars were well spent and, though many politicians did not like him, the people did.

In 1884 Cleveland was nominated for President by the Democrats. The Democrats had traditionally been the party of the South and had not won the presidency since the Civil War. But the Republicans were split, and many did not like their candidate, James Blaine from Maine. Cleveland won a close election and began to reform the federal government as he had done for Buffalo and for New York state as mayor and governor. He continued the reforms of the civil service. He forced the railroads to return millions of acres of government lands they had taken. And he made sure the navy was well-equipped for the least cost possible. Historians have commented that Cleveland worked harder than any other President. He stayed in his office working for long hours, studied every proposed bill carefully, and vetoed hundreds of bills that he believed were intended to dishonestly reward undeserving persons. He was warned that his actions might cost him re-election and he is said to have replied, "What's the use of being elected or re-elected unless you stand for something?"

One of the vexing issues that has frequently faced the nation is the question of tariffs. This was true as the 1888 election approached. A tariff on imported goods inevitably raises the prices of those goods. In 1888 this was felt to hurt farmers and persons of modest means. But it would allow American manufacturers to realize greater profits on their goods. Cleveland and the Democrats favored a low tariff so as not to hurt farmers and other workers. Though Cleveland won more popular votes, Benjamin Harrison was elected President by the electoral college.

1888				
Benjamin Harrison Levi P. Morton	Indiana New York	Republican	233	48%
Grover Cleveland Allen G. Thurman	New York Ohio	Democratic	168	49%

Benjamin Harrison

Benjamin Harrison lived from 1833 to 1901 and served as our 23rd president from 1889 to 1893. Though he spent most of his early years on his father's farm, he liked to read and was a good student. He graduated from Miami University in Oxford, Ohio and married by age twenty-one. He was pulled between the Presbyterian ministry and the law, and was known to carefully weigh alternatives and to allow no one to pressure him. He ultimately decided on law, and moved to Indianapolis to practice. In 1855 Harrison started a law practice with a partner. In 1857 he was elected Indianapolis city attorney and in 1860 was elected state supreme court reporter.

Harrison organized a regiment of volunteers when the Civil War began. His regiment was active in the fighting, and he is said to have been a brave and skilled officer. He won his soldiers' respect both by his courage and by his willingness to bandage his soldiers' wounds at the Battle of Kennesaw Mountain when "our surgeons got separated from us." He went back to his law practice after the war.

Harrison became a Senator in 1881. His name was well-known as his father had been a congressman and his grandfather, William Henry Harrison, had been elected as our ninth President. In 1886 Harrison lost his Senate seat. Though he did not seek the presidency in 1888, the convention was unable to settle on a candidate and, at the suggestion of industrialist Andrew Carnegie, Harrison was nominated.

Historians describe Harrison as having a public and a private persona. His private face was that of a kind man. But his public face was reserved and distracted. He was not a gregarious, hand-shaking politician. A friend once told him, "For God's sake, be human!" Harrison replied, "I tried it, but I failed. I'll never try it again." He ran what has been called a "front porch campaign," taking with small groups who came to see him.

Harrison did not win the popular vote but was elected by the electoral college. His goal was to continue the reform efforts of his predecessor, and he appointed Theodore Roosevelt commissioner to help in that effort. Historians regard Harrison as not having championed the Sherman Antitrust Act, though it was passed without his advocacy.

With Harrison's urging, congress passed the McKinley Tariff Act, which placed a high tariff on imports. Small businesses and farmers were put at a disadvantage by the resulting rise in prices of goods. In addition, big manufacturers were experiencing labor issues, as workers were forming unions and were protesting poor working conditions and child labor. Voters were disenchanted with Harrison.

Harrison lost the 1992 election, returned to Indianapolis, and had an active career. He practiced law, wrote, and lectured on constitutional law at Stanford University. British Guiana and Venezuela engaged in a border dispute and Harrison served the later as chief counsel. He practiced law until he died of pneumonia on March 13, 1901, at his home in Indianapolis. The poet James Whitcomb Riley gave a funeral eulogy, ad characterized Harrison as "a man both fearless and just."

1892				
Grover Cleveland Adlai E. Stevenson	New York Illinois	Democratic	277	46%
Benjamin Harrison Whitelaw Reid	Indiana New York	Republican	145	43%
James B. Weaver James G. Field	Iowa Virginia	People's (or Populist)	22	9%

Grover Cleveland

Grover Cleveland lived from 1837 to 1908. Having been our 22nd president from 1885 to 1889, he returned as our 24th president from 1893 to 1897. He was re-elected during a time of increased public awareness of the difficulties experienced by workers. Government has never truly had a laissez-faire policy toward the economy, and this was certainly true during Cleveland's second term. Railroad workers were on strike in Chicago, and Cleveland sent federal troops to break it. He is reported to have stated, "If it takes the entire army and navy to deliver one postcard in Chicago, that card will be delivered." This may have been admirable dedication to mail delivery, but it was not admirable support of the welfare and interests of workers.

The country has been subject to a series of economic downturns and depressions, and there was one at that time. Cleveland did not attempt to help people who were unemployed, as he did not believe such actions would have been constitutional. His efforts to improve the treasury system were not successful. History does note, however, that the federal government was a harder working and more honest one when he left office than when he entered. It appears Cleveland's first term as President was considerably more positive and successful than was his second.

Cleveland left the presidency in 1897 and moved to a spacious house in Princeton, New Jersey, where he practiced law, played the stock market, served on corporate boards and gave public speeches. He was treated like royalty by the town's inhabitants, and he became a trustee of Princeton University. He wrote essays and political commentary, including a book entitled *Presidential Problems*, which focused on some of his most controversial decisions. The death of his eldest daughter in 1904 visibly aged Cleveland, and he is said to have never fully recovered. He felt death nearing when he suffered a severe medical setback while on vacation in late March of 1908. He was rushed back to Princeton, where he died on June 24. It was reported that his last words were, "I have tried so hard to do right."

1896				
William McKinley Garret A. Hobart	Ohio New Jersey	Republican	271	51%
William Jennings Bryan Arthur Sewell	Nebraska Maine	Democratic / Populist	176	47%

1900				
William McKinley Theodore Roosevelt	Ohio New York	Republican	292	52%
William Jennings Bryan Adlai E. Stevenson	Nebraska Illinois	Democratic	155	46%

William McKinley

William McKinley lived from 1843 to 1901 and served as our 25th president from 1897 to 1901. McKinley was born and brought up in small towns in Ohio. Though not a gifted student he worked hard. He attended Allegheny College, but left after one year. There are varying accounts of why he left college, including illness, tight family finances, and dismissal due to a prank. He lived with his parents and taught school. He entered the Civil War as a private, and was promoted to major by the war's end. He then studied law and married. He reportedly told friends he had been convinced from an early age he would become President.

McKinley ran for Congress and served there for fourteen years, where he was a leader. He was elected governor of Ohio in 1991. In this era, powerful industrialists such as John D. Rockefeller, Andrew Carnegie and J.P. Morgan exerted more control over the nation's economy than did presidents. Though these industrialists competed fiercely with one another, they all feared a loss of their dominance if government were to enact policies limiting their power. They felt presidential candidate William Jennings Bryan was a threat to them, and they therefore sheathed their swords and worked together to have William McKinley nominated for the presidency. McKinley was essentially hand-picked by powerful industrialists for the purpose of keeping the government from limiting their wealth and dominance. With their help he was nominated in 1896. McKinley ran the same type of campaign as had Benjamin Harrison, a so-called "front porch campaign." Despite Democratic campaign claims that McKinley would be a pawn of rich bankers and industrialists, he was elected by a comfortable margin.

American expansion continued during McKinley's term. A major issue at the time concerned Cuba's fight for its independence from Spain. This was an era of what was termed "yellow journalism," a style of newspaper reporting favoring sensationalism over facts. Yellow journalism reached its peak, and newspapers fanned the flames of sentiment for war with

Spain. The newspapers seized on the sinking of an American ship, the Maine, in Havana Harbor. Though he apparently knew that Spain had finally agreed to grant Cuba its independence, McKinley nonetheless bowed to public sentiment and asked for a declaration of war on Spain. Acting Secretary of The Navy Theodore Roosevelt sent ships to the Philippines and destroyed Spain's naval fleet there. Three months later the War Department had assembled a strong enough military and the United States sent forces to Cuba. These forces defeated Spain, ending Spain's status as a major naval power. The U.S. took control of the Philippines, Guam and Puerto Rico. Hawaii also became an American territory during McKinley's term in office.

Business had recovered and McKinley was elected to a second term. However, he was shot and killed by a self-styled anarchist named Leon Czolgosc at a fair in Buffalo.

1904				
Theodore Roosevelt Charles W. Fairbanks	New York Indiana	Republican	336	56%
Alton B. Parker Henry G. Davis	New York West Virginia	Democratic	140	38%

Theodore Roosevelt

Theodore Roosevelt lived from 1858 to 1919 and served as our 26th president from 1901 to 1909. He was one of the most colorful and popular of all U.S. Presidents. As a child he was ill with bronchial asthma, and was sickly and frail. He read a lot about the exploits of adventuresome persons, and wanted to be like them. His father once told him that he had the mind but not the body to succeed, and said, "you must make your body work."

Theodore embarked on a devoted regimen of physical fitness. In addition to gymnasium workouts, he took to horseback riding and swimming, and he became particularly robust and strong. He did not neglect his mind. With the help of tutors, he learned English, French, German and Latin. At age eighteen he entered Harvard University, where he was known to be particularly energetic. He joined many clubs, rowed, boxed, and served on the editorial board of the undergraduate magazine. At age twenty-two he graduated and married.

Roosevelt entered law school, but left due to his distaste for what he saw as the greedy values of corporate law. Though he had been surrounded by wealth since birth, he was civic minded and detested corruption. He was advised by friends not to go into politics, which they considered "a dirty business." Roosevelt believed, however, that a person could transcend the dirt and work honestly for the country's benefit. He was elected to the state legislature at age twenty-three.

When Roosevelt was twenty-six his wife and his mother died within hours of one another, and he commented, "the light went out from my life." He dealt with the loss by moving to the Dakota territory, where he worked as a cowboy, branding steers and stopping stampedes. He then became a deputy sheriff. After two years he moved to New York and to politics. From age thirty-one to thirty-six he served as a United States Civil Service Commissioner. He was a reformer, and exposed corruption in the New York Customhouse. He ensured that thousands of federal jobs that had previously been awarded by the patronage system would be earned by merit. He was appointed New York police commissioner at age thirty-six due to his reputation as a tireless reformer. Corruption and bribes had ruled the police department. Roosevelt fired the police chief and the inspector, and even walked the streets at night to combat lazy patrols. At age thirty-nine Roosevelt was appointed assistant secretary of the navy by President McKinley.

When war was declared on Spain, Roosevelt organized a cavalry regiment called the Rough Riders. During the war in Cuba, he led his regiment up San Juan Hill to capture a Spanish fort. He would later say, "San Juan was the greatest day of my life."

After the war, Roosevelt returned to New York and conducted a vigorous and successful campaign for governor. He opened his office to reporters and used the newspapers as a "bully pulpit" to more fully inform and rally the public. Corporations had been receiving special treatment by the state, and Roosevelt supported a bill to force the corporations to pay a fair share of their profits. Though opposed by political bosses, Roosevelt managed to get the bill passed in what has been termed one of the greatest victories for reform in New York history.

Roosevelt was nominated for the vice presidency in 1990, the year the Sherman Anti-Trust Act was passed. When McKinley was assassinated, he succeeded him as President. Roosevelt believed in a strong and united country, and was fearful of the division between wealthy and poor. Businesses were able to defeat competition and pay low wages by forming trusts. The Progressive movement was growing and Roosevelt became its leader. Using the Sherman Anti-Trust Act, Roosevelt sued Northern Securities, a huge company that had established a monopoly in railroads. A federal court ordered Northern Securities dissolved, and the Supreme Court upheld the decision. Roosevelt did not promote "trust busting" per se, but wanted government to impose reasonable regulations on corporations.

Coal mines were a particularly difficult environment for workers. There was a strike in 1902, as workers wanted their union recognized, an 8-hour workday, and a ten-to-twenty percent raise in pay. Previous Presidents had often used federal troops to break strikes. In a major reversal of that tradition, Roosevelt threatened to use troops to regulate the mines. The owners agreed to submit the dispute to arbitration. The union was recognized and the work week was reduced. Workers received a ten percent raise and the companies were permitted to raise their prices by ten percent. Roosevelt stated his desire "to see fair play among all men, capitalists or wage-workers."

What may well have been the most significant legislation ever passed in the country was passed during Roosevelt's tenure: The Pure Food and Drug Act of 1906: "An act for preventing the manufacture, sale or transportation of adulterated, misbranded, poisonous or deleterious foods, drugs, medicines and liquors, for regulating traffic therein, and for other purposes."

Roosevelt was not liked by all politicians. He was enormously popular with the public and was re-elected handily in 1904. A major event of his presidency was the treaty that led to the building of the Panama Canal. The Isthmus of Panama belonged to Colombia, who refused to sell it to Roosevelt. There was a faction in the isthmus who wanted the canal and who planned to rebel. Roosevelt sent warships and, when the rebellion began, U.S. sailors prevented the Colombians from putting down the rebellion. The United States recognized the newly formed Republic of Panama, who leased the land for the canal to the U.S.

Despite his popularity, Roosevelt did not run for a third term at that time. He hunted in Africa, toured Europe, and wrote and gave speeches. However, Roosevelt was unhappy with President Taft and ran again in 1912. He formed a new party – the Progressive Party. He was shot and wounded while speaking in Milwaukee. He completed his speech, went to a hospital and, when asked about his condition, replied, "I feel as strong as a bull moose." Thus, his party began to be known as the Bull Moose Party. Woodrow Wilson was elected President in 1912 as the Republican vote was split between Roosevelt and Taft.

Historians have debated Roosevelt's philosophy and intent. The consensus appears to be that Roosevelt wanted to preserve the capitalist system, but to moderate its excesses. He believed that "the greed and arrogance of rich men," along with corruption in business and politics, could cause discontent and promote socialist movements. He stated, "If on this new continent we merely build another country of great but unjustly divided material prosperity, we shall have done nothing."

Roosevelt continued to live a purposeful and adventurous life. After losing the 1912 election to Woodrow Wilson, Roosevelt and his son Kermit voyaged to the jungles of Brazil to explore the Rio da Dúvida (River of Doubt) in the Amazon region. During the seven-month, 15,000-mile expedition, Roosevelt contracted malaria and injured his leg. He returned to the United States and wrote scientific essays and history books.

Roosevelt promoted military preparedness when World War I broke out. He was disappointed by President Wilson's call for neutrality and, when the U.S. did enter the war in 1917, offered to organize a volunteer division. His request was turned down by the War Department. All four of Roosevelt's sons volunteered to fight. His youngest son Quentin was killed while flying a mission in Germany and, for the first time in his life, sadness overtook the once unconquerable warrior. He continued to tour the nation to promote war bonds, but with a less enthusiastic voice. Theodore Roosevelt died of a pulmonary embolism in his sleep on January 6, 1919 in his home in Oyster Bay, New York. One commentator said that death had to take him while he slept, else it would have had a fight on its hands.

Issues and Trends in Era 205

The Advent of the Grange

What was known as "the Granger movement" began with the efforts of a Department of Agriculture employee named Oliver Hudson Kelley. Kelley toured the South in 1866 and was distressed by the lack of sound agricultural practices. In 1867 Kelley started an organization with the goal of bringing farmers together for educational and social purposes. It was called the Patrons of Husbandry. By 1870 nine states had Granges. By the mid-1870s nearly every state had at least one Grange, and national membership reached close to 800,000.

As described in the following section, farmers were doing poorly in the late nineteenth century. Their plight worsened during the financial crisis of 1873, resulting in a surge in Grange membership. At state and national levels, Grangers supported reform minded groups such as the Greenback Party and the Populist Party, hoping to elect officials sympathetic to them.

Many rural communities in the United States still have a Grange Hall, and local Granges are a center of community life for many non-urban communities. There is a National Grange with headquarters in Washington, D.C. In 2005, the National Grange had a membership of 160,000, with organizations in 2,100 communities in 36 states. The organization holds annual conventions and publishes its proceedings. In 2006 the organization described its mission and how it strives to achieve it through fellowship, service, and legislation. The mission statement included the following:

> The Grange provides opportunities for individuals and families to develop to their highest potential in order to build stronger communities and states, as well as a stronger nation.

Today's Grange is a non-partisan organization that supports policies, not political parties or candidates. Though originally founded to serve the interests of farmers, the Grange has broadened its range to include a wide variety of issues, and all are welcome to join the Grange.

The Plight of Farmers and the Rise of the Populist Party

The term "populist" is used in very contradictory ways today, and when used it bears no resemblance whatsoever to what it meant in Era 205. Today, the term may be used either in admiration (a man who supports the people) or in derision (not having substance, but appealing to the least knowledgeable people). The populist movement and the Populist Party had very specific origins and goals.

In the late nineteenth century farmers were doing extremely poorly, both economically and in the eyes of others. Thomas Jefferson had called farmers "God's chosen people," but in the late nineteenth century the words "hayseed" and "hick" were in vogue, and indicated contempt rather than admiration.

There were many factors limiting the success of farmers in the late nineteenth century. More acres than ever before were under cultivation, and harvesting yields were up by a factor of about six since the advent of the McCormack reaper. Therefore, there was significant overproduction which, due to supply/demand, was drastically lowering the price of yields. Tariffs were resulting in higher prices for equipment. Farmers needed to use borrowed money for land, equipment, seeds and day-to-day necessities, and financial institutions were not providing favorable interest rates. There was currency deflation, which meant it was difficult to pay back loans. Most farmers were in perpetual debt, with little hope to recover. In addition, there was a degree of social isolation as more and more people left rural areas for industrial work in cities, and there was a loss of social status. All in all, it was a dismal time for the farmer.

There was no help forthcoming from the political world. Adam Smith's ideas of supply and demand and laissez-faire were still influential. In addition, there was a theory that wages would always adjust to subsistence level. These ideas were considered "immutable laws of economics," and it was therefore considered futile to try to combat those laws and to attempt to assist farmers (or unskilled industrial workers). Furthermore, the political parties were evenly matched, and elections depended on a few votes in about a half dozen states. Politicians relied on big city machines and corporate contributions, rather than on substantive public discussions of issues, to obtain the needed votes. (The problems we face in current day electioneering are not new to the world).

From the farmers' point of view, the political system was corrupt and there was no relief in sight. Finally, farmers gathered to form associations, and ultimately created the Populist Party. The Populist Party advanced what has been termed "one of the most comprehensive reform documents in American History." The Populists called for an end to government subsidies for corporations, a graduated income tax, a more accommodating stance toward labor unions, and specific reforms to return more power to the people. The latter included the referendum, which allows citizens to vote on a statute passed by the legislature, enabling the voters to enact or repeal the measure.

In the 1892 elections, the Populist Party's presidential candidate, James B. Weaver, received a million poplar votes and twenty-two electoral votes. Grover Cleveland was elected President, though many Populists did win races for the Senate, Congress, governorships and seats in state legislatures. Some of the energy and ideas of the Populists contributed to the rise of the Progressive Party, which will be described in the discussion of Era 206 (Roaring Twenties, Prohibition, The Great Depression, World War II). One key point, as stated in the first paragraph of this section, is that when the word "populist" is used today, it is all but meaningless and bears little resemblance to the actual Populist Movement of the 1880s and 1890s.

The Sherman Antitrust Act

The Sherman Antitrust Act is a landmark federal statute passed by Congress in 1890 under the presidency of Benjamin Harrison. If government regulators view a business's practices to be competitive, the practice is allowed under the Act. But the Act recommends that the federal government investigate and pursue trusts that exert non-competitive control over a market.

This act did not apply to arrangements in which one party conveys property to a trustee to hold for a beneficiary, such as to hold inheritances for the benefit of children. The law applies to trusts that combine large businesses to create a monopoly – to exert complete control over a market. In most countries outside the United States, antitrust law is known as "competition law."

The law attempts to prevent businesses from artificially raising prices by restricting trade or supply. Monopoly achieved solely by merit, called "innocent monopoly," is not prohibited by the Act. However, actions by a monopolist to artificially preserve that status, or nefarious dealings to create a monopoly, are not legal. The purpose of the Sherman Act is not to protect competitors from losing out to legitimately successful businesses. And its purpose is not to prevent businesses from gaining honest profits. The purpose is to preserve a competitive marketplace and to protect consumers from abuses.

Riding the Rails - the Era of the Hobo

The rise of a population of migrant workers probably spanned Eras 204 (Civil War, First Oil Well, Robber Barons) through 206. It is a theme usually forgotten but sometimes romanticized. The era is commemorated in older songs such as *The Wabash Cannonball* and *The Big Rock Candy Mountain* and has found its way into American folklore and film. The migrant workers were at times called "bums" but more often called "hobos." There is a romanticized view of the hobo: a man refusing to be tied down to a routine life and instead seeking freedom and adventure on the rails. This is evidenced by Roger Miller's 1964 hit song *King of the Road*. Miller's lyrics describe the life of a vagabond who is poor but revels in his freedom, whimsically calling himself "king of the road."

The reality was that life as a hobo was difficult and dangerous. The itinerant worker was poor and was far from home and support. Traveling in railroad boxcars was far from safe. Such a traveler could be trapped between cars, fall under the wheels trying to board, or could die of exposure in bad weather. In addition, railroad security staff, nicknamed "bulls," had a reputation for violence against trespassers.

It is unclear exactly when hobos appeared on the American railroading scene. After the Civil War in the 1860s, many discharged veterans may have hopped freight trains to return home. Others went west aboard freight trains seeking work on the American frontier. The reality is that many thousands of willing workers had no place in the post-Civil War economy. Historians estimate that hobos in the United States numbered about 500,000 in 1906 and 700,000 in 1911.

The number of hobos increased markedly to over a million during the Great Depression era of the 1930s. With no work and no prospects at home, many decided to travel for free by freight train to try their luck elsewhere. For a variety of reasons, the practice is less common today, though it exists in folklore and song.

Inventions in Era 205

The Development of Radio

As with many inventions that history attributes to a single inventor, many people were involved in the invention of radio. Key to this work was the discovery of the connection between electricity and magnetism. These investigations began around 1820 with the work of Hans Orsted, and continued with the work of André-Marie Ampère, Joseph Henry, and Michael Faraday.

These efforts culminated in a theory of electromagnetism developed by James Clerk Maxwell. Maxwell's inquiries took place in the 1860s and 1870s. He studied electric fields and magnetic fields, and noticed that they combine to create waves.

He described the relationship between electricity and magnetism mathematically by what are referred to as "Maxwell's Equations." We now know that electromagnetic waves are on a continuum from the shortest (gamma waves) to the longest (radio waves) and include a narrow window that we experience as visible light.

Maxwell's work motivated many people to experiment with wireless communication. Several inventors saw evidence of electromagnetic waves before they were proven to exist.

Heinrich Hertz performed the first unequivocal transmission of electromagnetic waves. Ironically, Hertz considered his results to be of little value.

After Hertz's work, many people tried to improve the transmission and detection of electromagnetic waves. Around the turn of the 20th century, Guglielmo Marconi developed the first apparatus for long distance radio communication. In December 1900, Canadian inventor Reginald Fessenden became the first person to make a public radio broadcast. By 1910 these various wireless systems had come to be referred to by the common name "radio."

Marconi's greatest achievement came on December 12, 1901, when he received a message sent from England to St. John's, Newfoundland. The transatlantic transmission won him worldwide fame. Ironically, detractors of the project were correct when they declared that radio waves would not follow the curvature of the earth, as Marconi believed. In fact, Marconi's transatlantic radio signal had been headed into space when it was reflected off the ionosphere and bounced back down to Canada. Much remained to be learned about the behavior of radio waves and the role of the atmosphere in radio transmissions. Marconi would continue to play a leading role in discoveries and innovations about radio during the next three decades.

The Wright Brothers' First Flight

Until the early years of the twentieth century, conventional wisdom held that heavier than air machines cannot fly. Orville and Wilbur Wright believed otherwise, and in 1896 the brothers began experimenting with flight at their bicycle shop in Dayton, Ohio. They began with gliders, and chose the beach at Kitty Hawk for their experiments because that location had constant wind that added lift to their craft. In 1902 they made more than 700 successful flights with their glider at Kitty Hawk. The brothers then turned their attention to powered flight. They needed a light and powerful engine. As no automobile manufacturer could supply one, they designed and built their own.

Their hard work, experimentation and innovation came together in December of 1903. On December 17, 1903 Orville Wright piloted the first flight, which lasted 12 seconds, covered 120 feet, and reached an altitude of 20 feet above the beach. Three more flights were made that day with Orville's brother Wilbur piloting the record flight lasting 59 seconds over 852 feet. The brothers had notified several newspapers prior to the event, but only one - the local journal - made mention of the event. Nonetheless, their dedication, innovative spirit and daring enabled them to take to the sky and to forever change the course of history.

Medical Advances in Era 205

<u>Understanding of and Eventual Vaccine for Yellow Fever</u>

An important contribution in the field of medicine was the discovery of the cause of yellow fever by U.S. Army surgeon Major Walter Reed. Yellow Fever was a devastating disease affecting persons living and working in Central America. During the Spanish-American war, combat-related injuries claimed fewer lives among American soldiers than did yellow fever, malaria, and other diseases. Yellow fever continued to ravage Cubans and the American occupation force after the war. In 1900, a commission was appointed by Army Surgeon General George M. Sternberg to discover the cause of the disease and to devise a preventative measure.

The Reed Commission began work in June 1900. It had been previously thought that yellow fever was spread by poor sanitation. But Reed eventually proved that the disease was spread by female mosquitoes of the species Aedes Aegypti, which carried the virus from person to person with their bites. It was not until 1937 that Max Theiler of the Rockefeller Foundation formulated an improved version of a vaccine that had first been created ten years earlier.

The Discovery of X-rays

Few scientific breakthroughs have had as immediate an impact as Wilhelm Conrad Roentgen's discovery of X-rays. This discovery instantly revolutionized both physics and medicine. The X-ray quickly emerged from the laboratory into widespread use. Within a year the use of X-rays for diagnosis and therapy was an established part of medicine.

Roentgen worked at the University of Würzburg, where he experimented with light and other emission phenomena. In November 1895, Roentgen noticed that when he shielded a cathode ray tube called a Crookes tube with heavy black cardboard, the green fluorescent light caused a platinum/barium screen nine feet away to glow. This was too far away to be a reaction to cathode rays as he understood them. He determined that the fluorescence was caused by invisible rays that passed through the opaque black paper wrapped around the tube. We now know them as electrons. He then found that this new type of ray could pass through the soft tissues of the body, leaving bones and metals visible.

Roentgen quickly executed careful experiments. In January 1896, he made his first public presentation and demonstration, and an attending anatomist proposed that the new discovery be named "Roentgen's Rays." News spread rapidly throughout the world, and studios opened to take "bone portraits," further fueling public interest and imagination. Though there were such frivolous reactions, the medical community quickly recognized the importance of Roentgen's discovery. By February 1896, X-rays found their way into clinical use in the U.S.

Roentgen was awarded the first Nobel Prize in physics in 1901 for his discovery. When asked about his thoughts at the moment of discovery, he replied, "I didn't think, I investigated." Roentgen is widely recognized as a brilliant and altruistic researcher who sought neither honors nor financial profits. He rejected a title of nobility and donated his Nobel Prize money to his university. He took out no patents on X-rays, wanting the world to freely benefit from his work. His altruism came at considerable personal cost. At the time of his death in 1923, Roentgen was nearly bankrupt from inflation following World War I.

What We Learned about the Universe in Era 205

This was a remarkably rich era in terms of humankind's expanding knowledge of the Universe we inhabit.

The Discovery of Electrons

The idea that matter is made up of infinitely small, simple, indivisible pieces is hardly new. The Greek thinker Democritus suggested the idea 2,000 years ago. However, the idea has only been taken seriously for the past 200 years, and it really took off in Era 205.

The first major discovery that set off modern atomic theory was that atoms are not in fact the smallest things that exist. J. J. Thompson discovered the electron in 1897. He measured the mass of cathode rays, showing they were made of particles, but were around 1800 times lighter than the lightest atom, hydrogen. Therefore, they were not atoms, but a new particle, the first subatomic particle to be discovered. He originally called them "corpuscles" but later named them electron. He also showed they were identical to particles given off by photoelectric and radioactive materials. It was quickly recognized that they are the particles that carry electric currents in metal wires, and carry the negative electric charge within atoms. Thompson was given the 1906 Nobel Prize in Physics for this work. He had overturned the belief that atoms are the ultimate, indivisible particles of matter. Thompson incorrectly postulated that low mass, negatively charged electrons are distributed evenly throughout a charged medium like plums in a plum pudding. This became known as the plum pudding model for the atom. The negatively charged electrons would balance out the positively charged medium so that each atom would be of neutral charge.

The Discovery of the Atomic Nucleus

Ernest Rutherford disproved Thompson's plum pudding idea and established that atoms have nuclei. His discovery came by accident during experiments from 1909-1911. He fired alpha particles at very thin sheets of gold foil. Alpha particles consist of two protons and two neutrons, and are about 8000 times as massive as an electron. Rutherford wanted to know how much the alpha particles were deflected from their original course when they passed through gold foil. Because alpha particles are positively charged and electrons are negatively charged, the electrons were expected to slightly alter the trajectory of the alpha particles. The experiment would be like hitting a golf ball through light shrubbery. If a golf ball were to hit a leaf, it would pass through. Because it is much denser than the leaves, it would hardly be affected. Rutherford expected the deflection to be small but sufficient to indicate how electrons are distributed throughout the atom.

Rutherford was surprised that most of the alpha particles passed through the gold foil as though they were passing through empty space. But a small number of the alpha particles were deflected right back from where they had come. It was as though most of the golf balls were passing right through trees, while a few balls might hit a tree trunk and be bounced back.

This result shows that the mass of an atom is not as evenly distributed as Thompson had formerly assumed. Rutherford concluded that the mass of an atom is mostly concentrated in a nucleus made up of tightly bonded protons and neutrons, which are then orbited by electrons.

Einstein and Special Relativity Theory

One of the greatest leaps in human knowledge occurred during this era. A Swiss patent clerk in 1905 rocked the world of physics when he published his special theory of relativity, which held that the laws of physics are the same for all observers *who are in the same frame of reference*, and that the speed of light in a vacuum is independent of the motion of all observers. Albert Einstein thus introduced a new framework for all of physics and proposed new concepts of space and time.

Among the key concepts that arose from Einstein's theory of special relativity is the equivalence of mass and energy. In physics, mass/energy equivalence states that anything having mass has an equivalent amount of energy and vice versa, with these fundamental quantities directly relating to one another by Einstein's famous formula: $E = m\,c\text{-squared}$.

This formula states that the equivalent energy (E) can be calculated as the mass (m) multiplied by the speed of light (c = about 3×10^8 meters/second) squared. Because the speed of light is a very large number in everyday units, the formula implies that even an everyday object at rest with a modest amount of mass has a very large amount of intrinsic energy.

Einstein and General Relativity

Ever since the time of Sir Isaac Newton, gravity was seen as a force that attracts objects to one another. And, Newton's laws of gravitation appeared accurate. There was a small problem with the orbit of the planet Mercury around the Sun, but that appeared to be a minor detail.

Albert Einstein turned our understanding of the Universe on its head with his General Theory of Relativity. He argued that gravity is not a force per se, but is rather is an effect of the bending of four-dimensional space-time due to the presence of massive objects within it. Thus, the Moon does not travel in an elliptical orbit around the Earth, but rather travels in a straight line through curved space-time. That's easy enough to imagine, isn't it? No! It is not! We experience our world in three spatial dimensions and we cannot truly visualize four-dimensional space-time.

Nevertheless, Einstein's General Theory of Relativity has passed every test to which it has been put. Identical, coordinated atomic clocks on the Earth and in a plane flying at high altitude will tell time differently, as predicted by general relativity. During a Solar Eclipse, starlight coming from stars behind the Sun will be slightly bent toward the Sun, as predicted by general relativity. The orbit of the planet Mercury does not behave as Newtonian theory predicted, but it is consistent with the field equations of general relativity. And global positioning technology would not work without corrections for general relativity. Albert Einstein changed our world.

<u>There was an Age of Dinosaurs</u>

There is hardly a person alive who does not know that Earth was once the home of creatures called dinosaurs, and that there was a large and fearsome carnivorous dinosaur called Tyrannosaurus Rex. But 113 years had passed since the election of George Washington as President when the first T Rex was discovered. The discovery was made by Barnum Brown, who led an expedition to the Hell Creek Formation of Southeastern Montana. There, in 1902, he discovered and excavated the first documented remains of Tyrannosaurus Rex. The honor of naming the new find was left to Brown's boss, Henry Osborne, curator of the American Museum in New York. Osborne named the animal Dynamosaurus Imperiosis, or "Imperial, powerful lizard." The next year Brown dug up another skeleton, and Osborne named it Tyrannosaurus Rex, or "terrible lizard king." After some consideration, they realized the two were the same animal, and the Dynamosaurus name was dropped (Good thing! Otherwise, instead of saying "T Rex" we would have to say "D Imp", which does not have the same ring to it).

The discoveries in Montana, Wyoming, North Dakota and other locations did more than inform us of the prior existence of T Rex. We learned that what is now the American Northwest was once a lush and varied landscape, with high canopy forests, extensive, shallow inland seas, and herds of herbivorous dinosaurs such as the ceratopsians and the duckbills. Ultimately, paleontologists and geologists have pieced together a stunning record of Earth's past, a past that had not been revealed during the first half of the existence of the United States as an independent nation.

Human Origins

The history of the discovery and interpretation of fossil evidence of human origins is not a smooth one. Interpretation of fossil remains is difficult, and theories have been revised frequently. In 1891 Dutch surgeon Eugene Dubois found the first Homo Erectus ("upright man") fossil in Indonesia. The fossil is a one-million-year-old skull. Other Homo Erectus fossils have been found in Olduvai Gorge, Tanzania, a location where many human origins discoveries have been made. Homo Erectus is believed to have lived from about two million until perhaps three hundred thousand years ago. Many important discoveries have been made since then, but Dubois' discovery was among those that encouraged more research.

The Restless Earth in Era 205

The 1906 San Francisco Earthquake

On the morning of April 18, 1906, a massive earthquake shook San Francisco, California. The 1906 San Francisco Earthquake was the result of the interaction of two sections of the Earth's crust, namely the Pacific Tectonic Plate and the North American Plate on the boundary line that is now known as the San Andreas Fault. The Pacific Plate moves northward relative to the North American Plate, creating what is known as a "strike-slip fault." However, it would be more than thirty years after the 1906 earthquake before plate tectonics would be understood and would become the grand unifying theory of earth science.

The quake itself lasted less than a minute. But the immediate impact was disastrous and the disruption lasted months. The earthquake ignited several fires around the city. They burned for three days and destroyed nearly 500 city blocks.

San Francisco's large military contingent responded quickly but the city was devastated. Three thousand people died in the earthquake and fires, and half of the city's 400,000 residents were left homeless. The country and the world responded with aid, but those who survived faced weeks of difficulty and hardship.

The survivors slept in tents in city parks, and stood in long lines for food. The San Francisco earthquake is considered one of the worst natural disasters in U.S. history. Though it may have released less total energy than had been released in the New Madrid earthquakes of 1811-1812, it devastated a highly populated city.

The Tunguska Explosion

On June 30, 1908 there was a huge explosion near the Tunguska River in Siberia in the northern USSR. The blast levelled trees in an area of approximately 1500 square miles. Later expeditions to this remote area failed to locate a crater, which was a mystery, as the explosion was believed to have been caused by an impact. The incident started a series of celebrated rumors and legends, as some people speculated that the blast had been caused by an alien spaceship and others opined it had been caused by a tiny piece of antimatter.

It is now known that large meteors or small asteroids will often, when they pass through Earth's atmosphere, explode above the Earth's surface. It is estimated that the meteor that caused the Tunguska event measured between 60 and 200 meters in diameter. The explosion is estimated to have had the power of between 10 and 15 megatons and was the largest explosion caused by a comet or meteor in recorded human history until the explosion over Russia of the Chelyabinsk meteor in 2013. We now know that there are thousands of so-called "near-Earth objects" in orbit about the Sun, many of which could have Earth-crossing orbits. The largest of them have been detected and their orbits have been calculated. The search continues.

Miscellaneous Events in Era 205

United Fruit Company and American Imperialism

We have a tendency to believe that U.S. power in the world is wielded only by the government. And, we tend to view the United States as being benign and non-imperialistic. However, American companies whose activities cross national borders have been extremely powerful in the countries in which they do business. This has been viewed as American imperialism. The United Fruit Company is one such example.

The United Fruit Company was an American corporation that sold tropical fruit in the United States and Europe. The fruit was grown on Central and South American plantations. The company flourished in the early and mid-20th century, controlling vast territories and transportation networks in Central America, Colombia, Ecuador, and the West Indies. Though it did have competition (Standard Fruit Company, which later became Dole Food Company), it maintained a virtual monopoly in certain regions, some of which came to be called "banana republics," such as Costa Rica, Honduras, and Guatemala.

United Fruit had a deep and long-lasting impact on the economic and political development of several Latin American countries. Critics often accused it of exploitation and described it as the prime example of a multinational corporation intruding on the internal politics of another country. After a period of decline, United Fruit merged with AMK to become the United Brands Company, which in turn was transformed into Chiquita Brands International.

The intention is not to point the finger solely at United Fruit Company, but to help us have a balanced view of the global impacts of our nation, which are often, but not always, benign.

Summary Comments on Era 205

Every era seems to have a greater number of significant events packed into it than the preceding era, and Era 205 is no exception. All our preconceived notions about the nature of the Universe and the physical world - of matter, energy, gravitation, of the history of life - were revolutionized during this era. Note that not all citizens were aware of these new ideas, but the revolutionary knowledge nonetheless existed. Through the work of J.J. Thompson, Ernest Rutherford and many others, we now knew that matter was made up of atoms, which included electrons moving about a dense nucleus. Roentgen's discovery of X-rays changed medical diagnostic procedures in a virtual instant. Albert Einstein's theories of Special Relativity and General Relativity transformed the field of physics and our understanding of the nature of the physical world.

The distribution of information had been sped up considerably by Morse's telegraph and Hoe's rotary printing press in previous eras, but the work of Marconi and others that gave us radio transmission made instantaneous communication possible without a physical transmission wire. This transformed the flow of information and the reporting of news. And transportation would eventually be forever changed by the Wright Brothers' innovation of flight.

As indicated in the discussion of Era 204 (Civil War, First Oil Well, Robber Barons), the "Gilded Age" made vast fortunes for a few and propelled the United States to its position as the world's leading industrial power. However, most citizens were severely disadvantaged, as the gap between rich and poor grew markedly. Farmers, particularly hard hit, organized into groups that led to the Populist movement. Though the Populist movement did not achieve legislative victories, it did lead to the Progressive era, and in Era 206 (Roaring Twenties, Prohibition, The Great Depression, World War II) that movement would achieve legislative reform.

Powerful industrialists such as Rockefeller, Carnegie, Mellon, Hearst, and Ford had enormous impacts on the American scene, perhaps more than did Presidents Grover Cleveland, William Henry Harrison and William McKinley. Cleveland served as the country's 22nd and 24th President, the only person to lose and regain the presidency. Cleveland may have been the hardest working of all Presidents and, as he stated in his final words, he tried to do what was right. He fought against the use of public funds to funnel wealth to undeserving persons and he forced the railroads to give back lands that had been granted to them via preferential treatment. Nonetheless he continued a trend of using force to break strikes, as well as the trend of resisting providing aid to the needy on the basis that it would be unconstitutional.

Benjamin Harrison's presidency is noteworthy in that the Sherman Antitrust Act was signed into law. Under President William McKinley, Secretary of the Navy Theodore Roosevelt took military action that defeated Spain, giving the U.S. control of the Philippines, Guam and Puerto Rico. Hawaii also became an American territory via the same action.

Theodore Roosevelt's presidency was considerably more significant than those of his three predecessors. Roosevelt identified with the Progressive movement. He saw the discrepancy between rich and poor and he worked to limit the power of the trusts. Perhaps the most significant event of

Roosevelt's presidency was passage of the Pure Food and Drug Act. This may have marked the beginning of the government asserting its role of safeguarding the public interest over the interest of corporations.

The 1906 San Francisco earthquake and resulting fires devastated one of the west coast's most vibrant cities. The Tunguska explosion of 1908 occurred in a remote, unpopulated apart of Siberia. Though it did not result in many deaths, it did lead to decades of wild speculation until its cause was understood.

ERA: 206

Years: 1909 – 1933

(The Roaring Twenties, Prohibition, The Great Depression, World War II)

Presidents: William Howard Taft, Woodrow Wilson, Warren Harding, Calvin Coolidge and Herbert Hoover

1908				
William Howard Taft James S. Sherman	Ohio New York	Republican	321	52%
William Jennings Bryan John W. Kern	Nebraska Indiana	Democratic	162	43%

William Howard Taft

William Howard Taft lived from 1857 to 1930 and served as our 27th president from 1909 to 1913. His parents were strong figures. His mother, Luisa Torrey Taft, started a book club, an art association, and promoted kindergarten in public schools. His father, Alphonso, was a lawyer with a strong work ethic. Alphonso told his son that if he did his best on an exam and scored fifty, he would be sympathetic. But if he were to do less than his best and attain ninety-five, he would be ashamed of him.

William graduated second in his class at Yale, then studied law at his father's firm and at the Cincinnati Law School. He served as an assistant prosecutor and was appointed by President Chester Arthur as district internal revenue collector. He left the latter position after a year due to a dispute with Arthur.

In 1886 Taft married Helen Heron, who is said to have had a lifelong ambition to be First Lady. Taft was more interested in practicing law than being active in politics. In 1887 Taft was appointed to the Ohio Superior Court, and he felt this was the career for which he was best suited. His life's ambition was to be Chief Justice of the United States Supreme Court. Although he ultimately attained that goal, he would first serve as a somewhat reluctant President. Taft had the connections to be in politics, as his father was active in Republican politics, and his wife's father had been a law partner of Rutherford Hayes.

In 1892 President Harrison appointed Taft to a federal judgeship covering four states. As a federal judge, Taft's attitudes toward labor appear to have been inconsistent. During the Pullman strike in Chicago, he is said to have expressed in private letters the hope that federal troops would crush the workers. He is reported to have also written that if troops killed only six workers, it would hardly have made an impression. Paradoxically, he favored a union leader in the Pullman case, and said, "It is of benefit to the public that laborers should unite in their common interest and for lawful purposes. They have labor to sell." Taft was the first judge to argue in favor of workers' right to strike. It may have been that Taft favored "a peaceable strike," but railed against what he considered to be mob action.

The Philippine Islands had recently been annexed by the U.S., and President McKinley chose Taft to head a commission to govern. McKinley charged Taft with guiding the islands to peace, and to getting the Filipinos accustomed to legislative government. Taft wanted an island economy that would be of material benefit to the Filipinos. He also worked to expand public services and schools. The U.S. would at the same time have access to raw materials. Taft was sincere in his efforts, and was puzzled and angry when accused by a reporter of insincerity. Overall, he is said to have been well-liked there. Though service on the Supreme Court was Taft's goal, he twice refused an appointment as he felt he had more to accomplish on the islands

President Theodore Roosevelt appointed Taft to be Secretary of War and expressed confidence in him. Roosevelt promoted Taft's candidacy for President, saying "Taft will do all in his power to further the great causes for which I have fought." Taft's ambition was the Supreme Court, but he accepted the nomination for the presidency and was elected. He disappointed Progressives, first by appointing five corporate lawyers to his cabinet. Taft's rationale was that these legal minds were needed to carry out reforms while at the same time not injuring business. Taft favored a lower tariff, and entrusted Congress with providing one. He may have lacked the political acumen to convey his wishes or support them, and a bill to raise the tariff was passed. Progressives objected, and Taft promised to veto the bill, but he then changed his mind, later declaring that the bill was the best tariff bill ever passed. Progressives never forgave him.

The public showed disappointment in Taft when, in 1910, a Democratic majority was voted into Congress. Taft's anxiety over his failures was manifested in overeating. He had always been big, and had had nicknames such as "Big Bill." But in 1910 he is reported to have ballooned to 355 pounds and to have fallen asleep at inopportune moments. Despite starting the year 1910 on a poor note, he did have some accomplishments later in the year. He supported the Mann-Elkins Bill, written by Progressives, which gave the Interstate Commerce Commission the ability to suspend railroad rate increases for ten months in order to review them. And he also asked Congress to create a Commission on Economy and Efficiency, which paved the way for a budget office.

Taft put aside government-owned lands rich in coal and oil. He believed the people, and not private business, should profit from those resources. Some historians state Taft did a fair job of carrying out Roosevelt's progressive policies, but lacked the charisma to be recognized for his achievements. Taft was responsible and tried to be fair. He admired a statement by Lincoln about always doing his best, and had that quotation placed on his desk. Nonetheless Roosevelt and his followers felt Taft had not done enough to further progressive causes and were disenchanted. Roosevelt ran against him, and Woodrow Wilson won the presidency over the divided Republican Party.

It was reported that as Taft left the White House he said, "I am glad to be going. This is the lonesomest place in the world." After his presidency, Taft taught at Yale University Law School until President Warren G. Harding appointed him chief justice of the U.S. Supreme Court. Prior to this appointment, Taft served as co-chairman of the National Labor Board.

Taft served as chief justice until his death in 1930. Most of his decisions were considered "cautiously conservative" and constraining of government. The Clayton Anti-Trust Act had a provision that barred injunctions against labor picketing. Taft struck down that provision while deciding a case. Congress tried to discourage child labor via an excise tax on goods manufactured by children, but Taft ruled against their initiative. Those two decisions were neither labor-friendly nor child-friendly. Taft felt comfortable on the Supreme Court. His weight went down and he appeared to be happier. In 1925 he wrote, "The truth is that in my present life I don't remember that I ever was President."

On March 8, 1930, William Howard Taft died from complications of heart disease, high blood pressure, and inflammation of the bladder. His was the first presidential funeral broadcast on radio.

1912				
Woodrow Wilson Thomas R. Marshall	New Jersey Indiana	Democratic	435	42%
Theodore Roosevelt Hiram Johnson	New York California	Progressive	88	27%
William Howard Taft James S. Sherman	Ohio New York	Republican	8	23%

1916				
Woodrow Wilson Thomas R. Marshall	New Jersey Indiana	Democratic	277	49%
Charles Evans Hughes Charles W. Fairbanks	New York Indiana	Republican	254	46%

Woodrow Wilson

Woodrow Wilson lived from 1856 to 1924 and served as our 28th president from 1913 to 1921. Wilson may have been one of the most intelligent men to ascend to the presidency, and he was committed to his ideals. There is a story that when Woodrow was a child, a cousin saw him looking at a portrait. When his cousin asked the identity of the portrait, Woodrow is said to have replied, "That is the greatest statesman who ever lived – Gladstone – and when grow up to be a man I mean to be a great statesman too." (William Gladstone had served as British Prime Minister for twelve years).

Wilson was born in Virginia. His family moved to Augusta, Georgia when he was a year old, and to Columbia, South Carolina when he was fourteen. At age seventeen he attended Davidson College for a year but withdrew due to health concerns. At age eighteen he resumed his studies at Princeton. He excelled at Princeton, reading extensively about government and the lives of statesmen. He formed a debating club and wrote about political subjects. As a senior he wrote an essay in which he recommended that Congress enliven its debates by giving nonvoting seats to cabinet members. His essay won him acclaim when it was published in the *International Review*.

Wilson studied law at the University of Virginia. He attracted a standing-room only crowd when he delivered a speech about English politician John Bright. He wrote an article about

Gladstone in the college magazine, in which he stated, "Great men need a moral quality to distinguish themselves from the commonplace." Though he withdrew from the University due to illness and did not receive a law degree, he studied on his own. He was admitted to the bar at age twenty-six and opened a practice in Atlanta. He hated the process of currying favor to obtain legal business, and his passion was to accomplish something great - to reshape political thought.

Woodrow enrolled at Johns Hopkins University, and his book *Congressional Government* received strong reviews. His book was later accepted as his thesis and he was awarded his Ph.D. He married Ellen Louise Axson in 1885 and taught at Bryn Mawr College in Pennsylvania and Wesleyan University in Connecticut before being appointed full professor of jurisprudence and political economy at Princeton. By 1902 Woodrow had published five books. During these years, his views were regarded as conservative, as he supported states' rights and the gold standard, criticized labor protests, and opposed restrictions on corporations. He had turned down the college presidency at the University of Virginia three times when, in 1902, he accepted his desired position of president of Princeton University.

At Princeton, Wilson embarked on a program of reforms aimed at improving academic rigor and independent thought. He set a precedent later adopted by almost all colleges and universities. Students would spend their first two years in a common program before focusing on a major discipline during their last two years. He continued to teach political science and history, and his conservative views began to change. He was influenced by Theodore Roosevelt's progressive programs. Though originally protective of trusts, by 1910 he began to condemn excessive concentrations of wealth. In 1910 Wilson ran for governor of New Jersey, supported by political bosses who believed they could control him. He declared that he would enter office "with absolutely no pledges of any kind to prevent me from serving the people." Wilson now presented himself as a Progressive and, in 1911, the New Jersey legislature approved his entire reform program. Included were provisions providing workmen's compensation and giving the Public Service Commission the power to set utility rates.

The 1912 Democratic convention lasted seven days and required 43 ballots, and Wilson won the nomination for the presidency. He won the election easily over a divided Republican party, receiving 435 electoral votes to Roosevelt's 88 and Taft's 8.

Wilson believed, as had Theodore Roosevelt, that the role of the President is to represent and serve all the people. He saw Congress as serving the needs of special interests. Tariffs had long been an issue, and high tariffs favor manufacturers, raise prices and hurt farmers, workers and small businesses. Wilson saw tariffs as an example of special privilege, and he managed to convince Congress to lower the tariff. In addition, prior administrations had favored business over labor, sometimes in violent and oppressive ways. Wilson managed to get Congress to assert that it is not illegal for workers to go on strike. This was a major change in the relationship between workers and managements.

When Wilson assumed the presidency, the country lacked a coordinated banking system. Every fifteen or twenty years the country experienced financial panics and recessions, typically brought about by excesses of one type or another. The banking system was not able to cope with and moderate these events. As will be described in a later section, six men held a very secret meeting on Jekyll Island, Georgia in November of 1910. They developed a plan, calling for a reserve association. Progressive congressmen wrote the Owen-Glass Act creating the Federal Reserve Board, and Wilson backed the bill. The Federal Reserve Board continues to function in the present, and has great influence over the nation's economy.

Wilson's progressivism appeared to have left him when it came to race relations. For instance, when postmaster general Albert Burleson segregated his department and allowed thirty-five

African Americans to be fired in Atlanta, Wilson did not intervene.

Wilson's wife Ellen died of kidney disease in 1914, and Wilson went into a depression. In 1915 he married Edith Galt in a quiet ceremony.

World War I broke out when Wilson had been in office two years. He tried to help warring parties resolve their differences. In the campaign of 1916, the slogan "He kept us out of war" helped his re-election. However, it had become increasingly difficult to maintain neutrality. Germany invaded Belgium, in violation of a treaty. Wilson took a moderate stand and continued to pursue diplomatic negotiations when in 1915 a German submarine sank the British ship Lusitania, causing 1200 deaths. Finally, when German submarines began to sink American ships in the Atlantic Ocean, Wilson asked Congress to declare war on Germany, stating "The world must be made safe for democracy." In the summer of 1918, Allied forces secured victory.

Wilson sailed to Europe to lead the American delegation in the peace conference. He was greeted with adulation by Europeans, who saw him as a savior. He is famous for having formulated a Fourteen Point peace plan, which he had been preparing even before the hostilities ended. But the Allies sought revenge and reparations, and Wilson was obliged to surrender most of his fourteen points in order to save what he saw as the most important point: the establishment of a League of Nations. The League was formed after Germany's surrender in November of 1918. But the League would not succeed without the United States as a member, and membership required Senate approval. Some senators wanted changes to the plan, but Wilson wanted the plan passed as it had been written. Wilson then had a stroke and was partially paralyzed. For two months his wife Edith Bolling Galt Wilson was his link to the outside, which angered some members of Congress. Without Wilson's leadership, the Senate voted against U.S. membership and the League was ineffective. Wilson recovered to a degree, but was never truly well again and withdrew from politics.

His ideals, and his avowed hope for a world without war, cause many to refer to him as he wished to be known: a "statesman." In some ways he may have provided inspiration for subsequent politicians. However, there are some dark marks on his record. He failed to promote civil rights and racial equality. Further, the Alien and Sedition Acts of 1917 and 1918, intended to help ensure success in World War I, gave tools of oppression to government officials.

Wilson retired to his recently purchased home in Washington, DC, and formed a law partnership with his former secretary of state, Bainbridge Colby. But the partnership was dissolved when Wilson, who had never truly recovered from his stroke, was unable to do the work. Wilson died on February 3, 1924.

1920				
Warren G. Harding Calvin Coolidge	Ohio Massachusetts	Republican	404	61%
James M. Cox Franklin D. Roosevelt	Ohio New York	Democratic	127	35%

Warren G. Harding

Warren G. Harding lived from 1865 to 1923 and served as our 29th president from 1921 to 1923. Harding was a politician whose ability to win the favor of others may have been greater than his capacity for envisioning and planning for the common good. Harding was brought up in small towns in Ohio and attended small rural schools. His parents hoped he would become a minister, and sent him to Ohio Central College. He was involved there, serving as yearbook editor and engaging in some debates. He graduated in 1882 and then taught in a country school. He drifted for a while, found the study of law to be boring, and then worked at a newspaper called *The Democratic Mirror* in Marion, Ohio. Subsequently, he and friends purchased a failed newspaper called *The Star*. Along with *the Daily Star* Harding printed an openly Republican paper called *The Weekly Star*. Harding used *The Weekly Star* as a platform from which to unleash venom on his opponents. Harding was pursued vigorously by Florence Kling DeWolfe, a widow and the daughter of a local banker. It was considered an unlikely match. Florence is said to have dominated her husband. The newspaper was not doing well, as Harding preferred playing poker to working. Florence took over the newspaper and re-vitalized it.

Harding ran for county auditor in 1892, and enjoyed public speaking and campaigning. In 1898 he succeeded in being elected to the state senate. He is said to have enjoyed being away from home, playing poker, and socializing. Nonetheless a politician and lobbyist named Harry Dougherty, noting Harding's good looks and oratory skills, took on the project of making him President. Dougherty convinced Harding to run for lieutenant governor in 1902. Harding won, but refused re-election. Dougherty ran Harding for governor in 1910. Though he lost, he came to the attention of William Howard Taft, who selected Harding to be a speaker at the 1912 Republican National Convention. Harding lost another bid for governor, but in 1914 Dougherty convinced him to make a bid for the U.S. Senate. This time he won, and he enjoyed being in the Senate. He went to the horse track, played golf and played poker with his friends. Historians indicate that Harding was more interested in securing jobs for friends than in issues, and he is reported to have missed half the roll calls in the senate. He made many friends, though no one saw him as having much ability.

Dougherty arranged to have Harding give the keynote speech as the 1916 Republican National Convention. As the 1920 election drew near, many Americans, tired of war and of the strife associated with industry, longed for simpler times. Harding considered himself to lack the talent needed to be President, and is reported to have confided to a friend, "The only thing I really worry about is that I might be nominated." Harding ran in three primaries, losing two by large margins. But primaries were not as influential as they are today. As had been the case in some prior nominating conventions, a split among leading candidates resulted in a compromise candidate being nominated. That was the case in 1920 and, with Dougherty's assistance, Harding was given the nomination.

Though World War I had been over for two years, the public was tired of wartime issues and was leery of Europe. Harding's campaign pledge – "return to normalcy "– won him the election.

Harding set his sights on an accomplishment of merit: being a force for world peace. His Secretary of State, Charles Hughes, devised a plan for limiting the size of the world's navies. Nine countries participated in a three-month disarmament conference known as the Washington Naval Conference. Treaties were enacted that may have helped preserve peace during the 1920s. This was Harding's most important effort. His efforts on the domestic front were varied. He did stand up for laborers by backing collective bargaining. And he promoted the 1921 Federal Highway Act, which allocated funds for a national highway system.

Harding's candidacy and presidency were characterized by corruption and scandals. He was known to have had extramarital affairs and, in at least one case, hush money was paid to a woman to silence her during the presidential campaign. While in the White House, one of his affairs was with his secretary, Grace Cole. Unfortunately, many of Harding's appointees were old friends from Ohio, prompting some to refer to his administration as "The Ohio Gang." It was the Grant administration all over again, with appointed government officials using their positions for personal gain. Harding may or may not have known about his friends' nefarious actions, but in either case he did not stop it. In

August 1923 Harding became ill and died. Rumors of poisoning or suicide were never substantiated.

Stories of the scandals of Harding's administration came out after his death. The most famous of these was the Teapot Dome scandal. Teapot Dome in Wyoming was the site of navy petroleum reserves. There were other petroleum reserve sites involved, but Teapot Dome became the name by which the scandal was identified. Government lands were leased by Secretary of the Interior Albert Bacon Fall to private oil companies at low rates in exchange for bribes. Fall was convicted, but the scrutiny and prosecution of Harding's associates dd not end with Fall. Some members of the Ohio Gang went to prison and some committed suicide to avoid it. Harding was personally likeable and had a powerful political ally. But this does not appear to have made up for his lack of hard work or commitment to the common good.

1924				
Calvin Coolidge Charles G. Dawes	Massachusetts Illinois	Republican	382	54%
John W. Davis Charles W. Bryan	West Virginia Nebraska	Democratic	136	29%
Robert M. LaFollette Burton K. Wheeler	Wisconsin Montana	Progressive	13	17%

Calvin Coolidge

John Calvin Coolidge lived from 1872 to 1933 and served as our 30th president from 1923 to 1929. Coolidge was raised in Vermont. He grew up as a farm boy and his early education was in a one-room schoolhouse. He had been named after his father, and he used his middle name to distinguish himself from the elder Coolidge. When Calvin was thirteen years of age his mother died and he went to a private school called Black River Academy. At age nineteen he enrolled at Amherst College, where for three years he did not distinguish himself. As a senior he emerged socially and academically, and graduated *cum laude*.

After graduation he went to Northampton, Massachusetts and studied law at the law firm of Hammond and Field. Calvin's first foray into politics would be his campaigning on behalf of Hammond for district attorney and Field for mayor. In 1887 at the age of twenty-five he began his own law practice. His character and demeanor had disparate traits, as he was typically unexpressive and taciturn, but was also known for being witty and making wisecracks. He was also known to be both honest and ambitious.

By the age of twenty-seven he won two elections, first as town councilman and then as city solicitor. After taking time away from politics he ran for school board in 1905 and lost an election for the only time. He married Grace Goodhue that year, and by 1908 they had two sons. Grace at times confided to friends that she had thoughts about leaving him as he was very introverted and at times mean.

Coolidge won the mayor's office in 1910 and began to believe his political career could advance. He reduced the city's debt and at the same time raised teachers' pay. He was elected to the state senate in 1911 and became lieutenant governor in 1916 and governor in 1918. Due to his own modest background, in his early political career he identified with common people. He favored progressive reform and supported reducing the work week from seven days to six. As his career continued, however, a more repressive and less humanitarian side of him emerged.

Hardly anyone outside Massachusetts knew of Coolidge until 1919. Public attention was focused on him in that year. Wages were falling behind inflation, and there were labor strikes across the nation. Boston police suffered from poor working conditions and low pay, and they organized under the American Federation of Labor and went on strike. Boston Mayor Edwin Curtis contended that police work and union membership were incompatible, and he suspended the organizers. With most police on strike and robberies increasing, the mayor brought in the state guard, who fired into a crowd and killed two men. Coolidge subsequently called in additional state guardsmen to intimidate the striking policemen and quell the strike. He won national acclaim for moving against the strikers, and this action may have prompted the Republican Party to nominate Coolidge for Vice President in 1920. As Vice President, Coolidge was virtually invisible, and earned the moniker "Silent Cal." He became President in August 1923 when Harding died.

Coolidge was not a typical, gregarious, hand-shaking politician. His speeches were short and he lived up to his nickname. He may have been as close as any President to have followed a laissez-faire approach to domestic affairs and the economy. He stated, "The business of America is business," and he believed government should keep out of the way. History has shown that this was an extremely poor time for a laissez-faire approach. A few decades earlier, most Americans felt debt was something to be strictly avoided. However, during the "Roaring Twenties," the public had an increased taste for entertainment and material possessions, and few families were debt free. Business was booming and stock market prices were soaring. There was an enormous debt spiral building, as investors and speculators used borrowed money as collateral to borrow more money to invest in the market.

Coolidge supported the materialistic ethic and the extravagant urban/industrial economy. He wanted business to essentially run the country, and he wanted fewer government regulations. He did as little as possible himself. He said, "In the discharge of the duties of office there is one rule of action more important than all others. It consists of never doing anything that someone else can do for you."

Coolidge ascended to the presidency via the death of Warren Harding, but was elected on his own in 1924. As the election approached, his younger son died. Coolidge suffered depression and lost his energy for politics and his desire to deal with the responsibilities of his office. He had little interest in foreign affairs. He did want the U.S. to join the Court of International Justice, but the Senate rejected the initiative. In 1928 the U.S. and sixty other nations signed the Kellogg-Briand Pact outlawing war. The pact was merely a symbolic gesture, as there were no provisions for enforcement.

By 1926 there were signs of weakness in the economy. Brookings Institute research revealed that sixty percent of American families earned less than the two-thousand-dollar annual income that was considered sufficient for necessities. Herbert Hoover and others warned Coolidge about excessive stock speculation and irresponsible banking practices, but he ignored the warnings. In fact, he fanned the flames of over speculation by expressing his optimism about the stock market. Coolidge frequently told reporters that the economy was "a matter that I wouldn't happen to know anything about." Privately, Coolidge confided that he considered the market "rigged," but he refused to take action.

In 1928 he stated, "I do not choose to run." This was a fortunate decision for Coolidge and his place in history. Not everyone in the country was confident about the boom in business and in the stock market. There were those who warned that the boom would inevitably bring about a bust. Coolidge did not believe them and took no measures to regulate the rampant speculation in the markets.

A few months after Coolidge left office, the highly over-leveraged stock market crashed and the country abruptly descended into the worst economic depression in its history. Millions of people lost their jobs, homes and life savings. The suicide rate, which had been 12 suicides per 100,000 people, rose to 19 per 100,000 and stayed that high through the depression years. Although Coolidge's inaction contributed to the Great Depression, it was his successor who would bear the blame in the eyes of the public.

Upon retirement, Coolidge returned to Northampton, Massachusetts. He spent the next four years writing articles for national magazines, as well as his autobiography. For a year in 1931 the McClure Newspaper chain ran a syndicated column: "Thinking Things over with Calvin Coolidge." On January 5, 1933, just after lunch, Coolidge passed away from coronary thrombosis. He died in his bedroom, where he had gone to take his usual two-hour nap.

1928				
Herbert C. Hoover Charles Curtis	California Kansas	Republican	444	58%
Alfred E. Smith Joseph T. Robinson	New York Arkansas	Democratic	87	41%

Herbert Hoover

Herbert Hoover lived from 1874 to 1964. Born in West Branch, Iowa in 1874, Hoover was the first President born west of the Mississippi River. He served as our 31st president from 1929 to 1933. Hoover may be one of the most misunderstood and unappreciated presidents considering what he stood for and what he did in public life. His father, a blacksmith, died when Herbert was six years of age. When his mother died three years later, he went to live in Oregon with an aunt and uncle, who instilled in him the values of hard work and service to others. While working in his uncle's real estate office he overheard a man describing his travels as a mining engineer. Herbert was intrigued and at age seventeen went to Stanford University to study engineering. After graduating, he worked as a laborer in a mine. He worked seven ten-hour days per week for very little pay and under filthy conditions. He later stated, "I then learned what the bottom levels of real human despair are paved with."

At age twenty-three, Hoover was hired by London-based Bewick, Moreing & Company. He was sent to Australia to investigate a gold deposit. On his advice the company bought the mine, which became very profitable. He also went to China on the company's behalf, where he discovered coal deposits. Hoover was in the city of Tientsin when the Boxer Rebellion broke out, and he and his wife Lou helped defend the city. He continued to travel, and was instrumental in developing mineral deposits.

By age thirty-five Hoover had his own company, and by age thirty-nine he had amassed four million dollars. In 1900 Hoover wrote the book *Principles of Mining*. He expressed his idealism and humanitarianism when he wrote,

> The time when the employer could ride roughshod over his labor is disappearing with the doctrine of *laissez-faire* on which it was founded. The sooner this fact is recognized, the better for the employer.

Hoover was in London when World War I began. Many Americans traveling abroad were unable to get money to return home and Hoover intervened. He formed an organization to raise money to help those travelers, and Hoover himself donated much of the money for that cause. This was the first of several humanitarian acts that defined his life. Many Belgians had no access to food after their country had been captured by the Germans, and Hoover intervened again. He helped bring food from the United States past German lines and thus saved millions of persons from starvation.

President Wilson appointed Hoover to the post of food administrator when the United States entered the war. His role was to encourage Americans to save food, which would then be sent to Europe for our soldiers and allies. He succeeded despite having had no real authority. After World War I ended, Hoover did much more than bring food to Europe. He helped rebuild the European economy. His efforts included helping to restore communication, repair railroads and clear rivers. He was criticized by some for sending food to Germany, to which he replied, "The U.S. is not at war with German infants."

Hoover pursued the presidency in 1920. Warren Harding was nominated and elected, and Harding then appointed Hoover to be secretary of commerce. He revitalized what had been a moribund department, and it is said that his duties stretched into other areas of government as well. He sought participation from many sources, and relied on national conferences. Leaders were encouraged to provide facts, discuss issues and to try to reach consensus.

In 1921 there was a famine in Russia, and Hoover sent food. He heard a complaint about his providing aid for a communist regime, and he is said to have replied, "Twenty million people are starving. Whatever their politics, they shall be fed." In 1922 Hoover published his book *American Individualism*, in which he promoted civil liberty and scientific progress, and denounced *laissez-faire* economics. He managed relief efforts after the great Mississippi flood of 1927. Though he had never been on the ballot for any political office, Americans were aware of his humanitarian work. In 1928 he won the Republican nomination for President, and he was elected by a substantial majority.

Hoover promoted the Agricultural Marketing Act. This initiative helped farmers form cooperatives, and it gave a farm board the ability to enter the commodities market when needed to stabilize prices. Historians regard this as having been Hoover's best attainment as President.

162

Hoover's success and popularity took a sudden and dramatic downturn. On October 29, 1929 the stock market crashed and the country was thrown into a deep economic depression. A downward spiral ensued, as over-leveraged businesses failed, people were put out of work, more businesses failed, and banks became insolvent.

As his prior humanitarian efforts indicated, Hoover was a very caring individual. But he was among those who held conservative beliefs about the economy and about the role of government. He was a practitioner of and a staunch advocate of volunteerism, believing that help to persons in distress should be provided by churches and volunteer organizations. Initially, he felt that for government to provide relief to individuals or businesses would violate American principles. His point of view was supported by his Treasury Secretary, Andrew Mellon.

Hardly anyone understood the causes of the Great Depression. Since Hoover was at the wheel of the ship of state when that ship foundered, he was blamed. Many people who had lost their homes lived in makeshift villages of shacks, which were termed "Hoovervilles." This was unfair. Nonetheless, though Hoover had not caused the Depression, he was slow to act to alleviate the suffering it caused. He finally did act to lend government money to businesses. In December of 1930 he asked Congress to allocate one hundred million dollars for a program of Public Works, with the intent of employing people to improve rivers and harbors and to construct highways and buildings. He worked eighteen-hour days and, though he was slow to start, he ultimately took more emergency economic initiatives than had any prior President.

The causes of the Depression were too widely systemic, Hoover's efforts were too little and too late, and there was no noticeable improvement in the nation's economy. Although he had started his work life alongside workers in California's mine shafts, and had championed and personally supported many humanitarian causes, he lost the confidence of the public. Hoover was soundly defeated in the presidential election of 1932. He went back to private life, but his days as a humanitarian were not over. In 1940, Hoover raised large quantities of food to help the Finns, who were at war with the Soviet Union. After World War II President Truman chose Hoover to lead an Emergency Commission to send food to citizens of countries decimated by war. Hoover's failure to act quickly to combat the Great Depression was clearly not due to a lack of empathy for the suffering of people, but was rather due to his reluctance to abandon a conservative philosophy of noninterference by government in the economy. His unyielding devotion to that principle has probably resulted in a lack of appreciation for him as a great humanitarian.

Herbert Hoover was only fifty-eight years old upon his defeat in the election of 1932. Immediately after the inauguration of Franklin Roosevelt, Hoover returned home to Palo Alto, California. For decades the public and the Democratic Party blamed Hoover for the Great Depression. Few Republicans in the 1930s wanted Hoover involved in party politics because of his standing in the popular mind. But Hoover was still concerned with national affairs and, in letters and essays, he attacked Roosevelt's New Deal initiatives. In particular, he criticized Roosevelt's decisions to abandon the gold standard, to recognize the Soviet Union, to support government intervention in the economy, and to build the foundations of a welfare state. In the 1936 presidential race he supported Republican Alfred Landon, who lost to Roosevelt by a wide margin.

Hoover traveled extensively in his post-presidential years. In 1938 he met with Adolf Hitler. Reportedly, Hitler was shouting during their meeting, and Hoover dressed down the German dictator. Hoover was among those who opposed U.S. intervention in Europe after Germany invaded Poland in 1939. It was not until Japan's surprise attack on Pearl Harbor that he changed his mind. War created a need for Hoover's organizational and humanitarian skills, as it had in 1918. FDR put aside his resentment for Hoover's attacks on him and supported

Hoover's appointment to lead an international relief effort for Belgium, Finland and Poland. However, Hoover was unsuccessful in bringing food relief to nations occupied by the Nazis.

In the post-World War II years, Hoover remained committed to public service and to commenting on domestic and international affairs. In 1947, Congress named Hoover chairman of the Commission on the Organization of the Executive Branch of Government, which became known as the "Hoover Commission." Republicans hoped the commission would curtail FDR's New Deal policies. However, Hoover now seemingly recognized that a more populous and complex nation placed greater burdens on the presidency, and he favored reforms to strengthen the Executive Branch. The 1949 Executive Reorganization Act included several of the Hoover Commission's proposals.

Hoover's position on America's Cold War policies was at times supportive and at times critical. He agreed with President Truman's post WWII initiative to rebuild Germany economically and politically to combat Soviet expansion. But he disagreed with U.S. involvement in Asia. He advocated building American military power and basing American strategy on defense of the Western Hemisphere.

Shortly before his death in 1964, Hoover endorsed Senator Barry Goldwater for President, stating that the conservative Arizona Republican's views matched his own on limiting federal authority over everyday life and the economy. Herbert Hoover died of colon cancer on October 20, 1964, at the age of ninety.

Issues and Themes in Era 206

The Alien and Sedition Acts of 1917/1918

Not everyone in the country favored the U.S. entrance into World War I. Congress and President Wilson feared that anti-war sentiments could undermine the war effort, and they therefore passed two laws: the Espionage Act of 1917 and the Sedition Act of 1918. (These were different and separate from the Alien and Sedition Acts passed under President John Adams in 1798 that were mostly repealed or expired by 1802.)

Congress' primary purpose with the Espionage Act was to combat *actual espionage* on behalf of America's enemies, such as publishing secret U.S. military plans. However, Section 3 of the Espionage Act criminalized "any disloyal, profane, scurrilous, or abusive language" about the U.S. government or military, or any speech intended to "incite insubordination, disloyalty, mutiny, or refusal of duty" in the military. Anyone found guilty of such acts would be subject to a fine of $10,000 and a prison sentence of 20 years.

The Espionage Act was reinforced by the Sedition Act of the following year, which imposed similarly harsh penalties on anyone found guilty of insulting or abusing the U.S. government, the flag, the Constitution or the military. These acts were inimical to America's long-held protection of the right of dissent and of the free exchange of ideas. It is not likely that the exchange of ideas, and anti-war sentiment, would have impaired the war effort. But the Alien and Sedition Acts did give power to those who would use them to persecute Americans with views different from theirs.

Both pieces of legislation were aimed at socialists, pacifists and other anti-war activists during World War I. They were used as tools of oppression following the war, during a period characterized by fear of communist influence on American society. This became known as the first "Red Scare." (A second, and even more devastating, red scare would occur during the 1940s and 1950s, associated largely with Senator Joseph McCarthy). Attorney General Alexander Palmer and his right-hand man, J. Edgar Hoover, liberally employed the Espionage and Sedition Acts to persecute left-wing political figures.

The Dollar Replaces the Pound

"The sun never sets on the British Empire." That was the saying when Great Britain was at its height of power. But the British Empire overextended itself and declined in economic and military power. The United States supplanted Great Britain as the world's strongest economic power. Consider the table below, which displays the number of dollars it would take to purchase a British pound.

Year	Dollars per pound
1864	$9.97
1914	$4.94
1950	$2.80
1985	$1.29

By the end of World War I, the United States had become the world's leading power and the U.S. dollar replaced the British pound as the "world's reserve currency."

The Progressive Movement and Political Labels

In the late 1800's a political movement began and was known as "Progressivism." Although this movement had its roots in Era 205 (Telegraph, Populist Movement, Pure Food and Drug Act), it reached full bloom in Era 206. Presidents Teddy Roosevelt and Woodrow Wilson identified themselves as Progressives. There was a view that American political and economic life had been dominated by corruption and by abuses of power. Progressives intended to stop those abuses and to establish more fairness in government and business affairs. The movement was not a unified one, and it had many very contradictory components. Nonetheless there were some unifying factors, and the term is still in use today. Therefore, it is worth exploring.

Some of the most consistent thrusts of Progressivism were the following:

Governmental reform emphasizing efforts to eliminate
   corruption.

An emphasis on science and expertise as guiding factors in
   reform.

Fairness and consumer protection.

Assistance to persons victimized by excesses of industrialization,

such as children, women and immigrants.

Regulation of big business to maintain balance of power.

Education/standards for professions (doctors and lawyers).

Restoration of a sense of ethics and of service to the nation
above narrow self-interest.

A view of government as a solution to problems rather than as
the problem.

Those are lofty goals, but not all Progressives promoted policies
consistent with them. And, even when their goals were similar,
they often favored very different means to attain those goals.

In 1776 Adam Smith published *The Wealth of Nations*, in which
he favored a "laissez-faire" approach to economics. In French,
"laissez-faire" means "let to do," meaning government should
not regulate economic life, but rather should stand back and
allow the "unseen hand" of market forces regulate economic life
naturally. Laissez-faire has often been cited as a guiding
government action or inaction. But the United States has never
truly followed a laissez-faire policy. To the extent that it has, it
has failed. Without government intervention, business and
industry have exploited the working class. Calvin Coolidge's
failure to use governmental power to restrain unbridled financial
speculation helped lead to the Great Depression.

In 1912, Teddy Roosevelt and Woodrow Wilson, two avowed
Progressives, ran against each other. They agreed that
government should be used to regulate business. But Wilson

believed that big business had to be broken up to restore competition. Roosevelt believed government should not break up big business, but should control it and legitimize its practices.

The contradictions within the Progressive movement ran deeper than that. Although fairness and equal opportunity seemed to be woven into the fabric of Progressivism, some Progressives sought to limit immigration. Others sought to limit the participation of women in political and economic life to protect them from their own supposed inferiority. Some wanted to use the government to impose their ideas of morality on the public. And, the Espionage and Sedition Act of 1917/1918, which virtually outlawed criticism of the government during World War I, hardly passes muster to promote a fair, ethical national mood.

The Progressive era saw the passage of amendments sixteen through twenty to the Constitution. These amendments established the income tax, direct election of senators, alcohol prohibition and women's' right to vote. Prohibition clearly did not attain the goal of a more moral population, but rather encouraged otherwise law-abiding persons to violate the law, and ushered in a period of organized crime.

Despite its contradictions and problems, the Progressive era accomplished much. It established the income tax, direct election of senators, and women's voting rights. It led to the establishment of professional standards for professions. The Pure Food and Drug Act and antitrust legislation were enacted. Perhaps the most important legacy is the view that government can be utilized to solve problems and to help the citizenry.

Many people trace today's philosophy of liberalism to Progressivism, and there is some truth to that view. However, perhaps a look at the terms Progressive, liberal and conservative is in order. Are they truly meaningful, or are they misleading? Do they help to understand different positions, or do they lead to distortion, misunderstanding and polarization?

Let's take the word "conservative." Dictionary definitions:

1. disposed to preserve existing conditions, institutions, or to restore traditional ones, and to limit change.

2. cautiously moderate or purposefully low, such as a conservative estimate.

3. traditional in style or manner; avoiding novelty or showiness.

Definition #1 is generally seen as indicative of political conservatism. But, is it apt? Consider an example. Political conservatives almost always favor of less regulation, including less environmental regulation, allowing industry to proceed unfettered by environmental restriction. Is that conservative? It is not. It is axiomatic in the biological sciences that "You can never do one thing." Any change will have a cascade of effects that we cannot predict. Allowing industry free reign regarding environmental impact is not conservative; indeed, it is radical.

Another example is gun control. Political conservatives strongly oppose controls on guns, even in an era in which mass shootings have become common. Refusing to back measures that would protect people from gun violence is radical, not conservative.

The following are some dictionary definitions of "liberal":

1. favorable to progress or reform in political affairs.

2. advocating the freedom of the individual and governmental guarantees of individual rights and liberties.

3. of or relating to representational forms of government rather than aristocracies and monarchies.

4. free from prejudice or bigotry; tolerant.

Many people who regard themselves as staunch conservatives believe strongly in individual freedom. And reform itself is not ether a liberal or a conservative value; it is the type of reform that may be construed one way or the other. The problem is that the terms "liberal and conservative" cause people to be polarized and to be against one another. Perhaps a better approach would be to identify an existing problem and then to discuss the relative merits of alternative solutions. The labels themselves are almost meaningless in the absence of such a process.

<u>Women's Right to Vote</u>

The nineteenth amendment to the United States Constitution was passed in June of 1920 and ratified on August 18, 1920. This amendment granted all women the right to vote. This right was won through diligence and courage over a period of years. Some of the key figures in the fight were Elizabeth Cady Stanton, Susan B. Anthony and Lucy Stone. While today we take for granted women's voting rights, we must remember that our history as a civilization and as a country has been a continually evolving process. Women's rights are one of several issues whose histories are surprising by the values and standards of today.

<u>The Scopes Trial</u>

One recurring issue in American life is the disagreement between acceptance of what we learn from science and rejection of science in the name of religious fundamentalism. This was never clearer than in the Scopes Trial, sometimes known infamously as "the Scopes monkey trial." At issue in 1925 was Tennessee's recently enacted law making it illegal to teach the theory of evolution. Science teacher John Scopes was prosecuted for having allegedly taught evolution in a Tennessee public school. The trial featured two of the best-known orators of the era, William Jennings Bryan and Clarence Darrow, as opposing attorneys. The trial was seen as a way to challenge the constitutionality of the bill, to publicly advocate for the legitimacy of Darwin's theory of evolution, and to enhance the profile of the American Civil Liberties Union (ACLU).

Interestingly, the trial may have been the result of a publicity stunt for the town of Dayton, Tennessee. A local businessman reportedly met with the school superintendent and a lawyer to discuss using the ACLU for newspaper attention to the town. The group asked if high school science teacher John Scopes would admit to teaching evolution for the purposes of prosecution. Scopes was a math and physics teacher and, though he accepted evolution, was not sure he had ever actually taught it.

Three-time presidential nominee William Jennings Bryan volunteered to present the prosecution's case. Bryan was known as an anti-evolution activist, having created national controversy over the teaching of evolution. Clarence Darrow was the attorney for the defense. Darrow's goal was to refute fundamentalist Christianity and raise awareness of the narrowness of taking the Bible literally. Darrow and Bryan had already contested this issue in the press and public debates.

The trial was more of a circus than a trial. Darrow called Bryan to the stand as an expert witness, which was a shock to the court. Darrow interrogated Bryan on interpreting the Bible literally, which undercut his earlier sweeping religious speeches. Bryan had to admit that he did not know much about science since the Bible did not provide any answers.

The judge ruled Bryan's testimony be taken from the record, and Darrow then suggested his client be found guilty to save time. The jury took nine minutes to pronounce Scopes guilty. Scopes was fined $100 and was offered a new teaching contract. Instead, he chose to leave Dayton and attend graduate school at the University of Chicago to study geology. The Scopes trial revealed how emotion can overcome reason when religious fervor is involved.

The Birth of the Federal Reserve

In November 1910, six men had a very secret meeting at the Jekyll Island Club, on an island off the coast of Georgia. They were there to write a plan to reform the nation's banking system. The six men were Nelson Aldrich, A. Piatt Andrew, Henry Davison, Arthur Shelton, Frank Vanderlip and Paul Warburg. The meeting and its purpose were such a closely guarded secret that that it was not until the 1930s that participants admitted the meeting had occurred. The plan written on Jekyll Island was the foundation for the Federal Reserve System.

American financial life had been characterized by financial panics and recessions, which seemed to occur every fifteen to twenty years. These events were typically preceded by excesses of one type or another, and they had significantly disrupted economic activity during the nineteenth century. Andrew Jackson had destroyed the national bank a century earlier. The occasional panics caused financial institutions to suspend operations, and this triggered long and deep recessions. American banks were required to hold large reserves of cash, but these reserves were scattered throughout the nation. The funds became effectively frozen in place and could not be used to help alleviate the financial crises. In 1910 the six participants at Jekyll Island wrote a plan that addressed their concerns. This report would later be presented to the National Monetary Commission.

Their plan called for a Reserve Association of America. This would be a single central bank with fifteen branches across the country, with branches governed by boards of directors elected by the member banks in their district. The branches would hold the reserves of their member banks. They would be empowered to issue currency, clear and collect checks, and transfer balances between branches. The national body would set discount rates for the system as a whole and could buy and sell securities. These latter two functions would eventually lie at the heart of the Federal Reserve's power to influence the economy. The Federal Reserve Act was passed by Congress in 1913.

An Age of Entertainment

Wages in the 1920s were relatively high by historical standards. This was partly due to Henry Ford having set a new standard for wages (which he did to enable his workers to become buyers of his automobiles). Automobiles were being mass-produced and were affordable, so many Americans bought them. Gasoline was also low in cost, and America was on the move as never before.

There was a shift in the lifestyle of many Americans. Many persons moved from a rural, agrarian existence to an urban and suburban lifestyle, as business needed a new class of employees: accountants, bookkeepers and middle managers. Business was booming, and many Americans were developing a taste for a new, faster lifestyle. There was considerably more emphasis on entertainment. Many sources have referred to the 1920s as "The Roaring Twenties." Modern music became popular as a result of the advent of radio and phonograph records.

The movie industry rose strongly in popularity due to the 1927 advent of sound (movies were now referred to as "talkies"). By the early 1920's, many American towns had a movie theater. The movie industry became big business, and historians estimate that three-quarters of Americans visited a movie theater every week. People may not have known the names of government officials, but they knew the names of leading actors and actresses. Baseball had emerged as the national pastime. From 1920 to 1929, minor league teams had been added and baseball total attendance increased more than 1000 percent, from just over nine million to just under ninety-three million.

Automobiles helped Americans spread out geographically while radio helped bring them closer culturally. In 1920, Pittsburgh's KDKA was the country's first commercial radio station. By mid-decade there were more than 500 stations in the country, and by the end of the 1920s, 12 million households had radios. Large networks could broadcast the same radio and news program to many stations at the same time. Soon, many Americans in different locales listened to the same programs and had immediate access to news. They may have laughed at the same jokes, sung the same songs, and essentially shared a culture.

Overall, many Americans were less concerned with subsistence and more engaged in entertainment. And it was a new era of mass culture. By the end of Era 206, of course, much of this activity would be reduced due to the Great Depression. But it was certainly an era in which American culture, tastes and lifestyles underwent a permanent transformation.

<u>Prohibition and the Roaring Twenties</u>

As noted in the section above – "An Age of Entertainment" – the 1920s saw a shift to urban living, higher wages for workers, more middle management and professional occupations, greater mobility, and shared culture through radio and movies. In 1920, women won the right to vote via the nineteenth amendment to the Constitution. Many women discarded prior strict, Victorian ways of dress and behavior, and participated more in economic life. The most familiar symbol of the "Roaring Twenties" may be "the flapper." This refers to young women who wore short skirts (just below the knee), wore a short hairstyle known as bobbed hair and who felt free to behave in ways that may have previously been considered "unladylike." Undoubtedly, though most young women in the 1920s were not representative of the flapper image, they nonetheless did gain unprecedented freedom from it. Millions of women worked in white-collar office jobs and could participate in the growing consumer economy.

It was in the context of the roaring twenties that an interesting and disruptive event took place: prohibition. It is estimated that during the nineteenth century, alcohol consumption had been three times higher than today's level. Alcohol abuse no doubt did have deleterious effects on many persons' health, work and families. It was in this environment that attempts were made to control or eliminate alcohol use. In the early 19th century, groups such the American Temperance Society had campaigned vigorously against drunkenness. In 1851 the state of Maine legislature passed a statewide prohibition on selling alcohol. A dozen other states subsequently instituted their own "Maine Laws," but later repealed them due to widespread opposition.

Opponents of alcohol use ultimately influenced Congress. On January 29, 1919, Congress ratified the 18th amendment to the Constitution. This prohibited the manufacture, transportation and sale of alcohol within the United States. This amendment went into effect a year later in January 1920, and had profound effects on the social, political, health and legal life of the country during the period of 1920 to 1933.

Interestingly, drinking alcohol was not explicitly prohibited. By law, Americans were allowed to keep and enjoy any spirits they had accumulated by January 1920 in the privacy of their homes. Wine drinking for religious ceremonial practices was allowed. Moreover, drug stores were allowed to sell "medicinal whiskey" to treat conditions such as toothaches or flu. With a physician's prescription, "patients" could legally buy a pint of hard liquor every ten days. This pharmaceutical whiskey often came with peculiar doctor's orders such as "Take three ounces every hour for stimulation until stimulated." Prohibition did very little to reduce alcohol consumption, and certainly did nothing to increase respect for law.

Prohibition made criminals out of many formerly law-abiding citizens. People who would not have previously broken the law would often, in protest, find ways to defy the law and drink. And police, who may have upheld the law previously, were inveigled into unlawful activity, perhaps by the combination of disrespect for the law in addition to lucrative payoffs. A very common option was to go to private, unlicensed barrooms, nicknamed "speakeasies." There was a "cloak and dagger" aspect to this, as patrons would sometimes have to speak the "password" to gain entry. This trend caused a permanent shift in American social life. Speakeasies multiplied, especially in urban areas. Some were fancy clubs with bands and ballroom dance floors. Others were dingy backrooms or basements.

Criminality did not decline during this era, but rose significantly due to the illegal manufacture and trafficking of liquor. Distillers of banned liquor were known as "bootleggers." They produced millions of gallons of poor-quality liquor during Prohibition, and they imported enormous amounts of spirits from other nations. Organized crime exploded in America, as criminals seized the opportunity to exploit the new rackets of bootlegging and speakeasies. The struggle between organized crime and law enforcement has been memorialized in American lore.

Prohibition was not universally enforced. Some states did not enact enforcement codes, and others did not appropriate any funds for enforcement. Prohibition was sometimes known as "the noble experiment," but it hardly attained nobility. It reduced respect for law, encouraged organized crime, and was bad for the economy. Nine years into the Prohibition Era, the county fell into the Great Depression. It was clear that money would be spent on alcohol under any circumstance. The public agreed with the argument that money would be better going into tax revenues than into the hands of organized crime. During the 1932 presidential campaign, Franklin D. Roosevelt called for a repeal of Prohibition. A year later, most states ratified the 21st Amendment, which repealed the 18th Amendment and ended Prohibition. Roosevelt is said to have marked the occasion by drinking a dirty martini. Prohibition was over, but some of the cultural changes it fostered endured.

Break-up of Standard Oil

Standard Oil Co. was established by John D. Rockefeller and Henry Flagler in 1870. Standard Oil's controversial history as one of the world's first and largest multinational corporations ended in 1911, when the U.S. Supreme Court ruled that Standard Oil was an illegal monopoly.

The Sherman Anti-Trust Act was passed in 1890. In 1906 President Teddy Roosevelt's U.S. Attorney general filed suit against Standard Oil under the Act. Rockefeller initially took to the road to avoid being served, but ultimately faced the courts. Rockefeller was defiant and asserted that he had used the methods he did out of necessity. The court held that Standard's actions and secret transport deals with railroads had resulted in a drop in kerosene prices from 58 cents in 1865 to 26 cents 1870. Standard was found to have used aggressive pricing to form a monopoly and to destroy competitors.

Thus, in 1911, in a landmark case, the U.S. Supreme Court ruled that Standard Oil was an illegal monopoly and had to be broken up. The company was divided into thirty-four different entities (for example, ExxonMobil, Marathon Petroleum, Amoco, and Chevron are some familiar names). Many of these companies are still by themselves among the largest corporations in the world. Interestingly, the total value of Standard Oil's offspring was greater than that of the parent company and, as a shareholder of all the smaller companies, Rockefeller became the richest person in modern history. He spent his later years engaged in philanthropy.

Inventions that Changed Our Lives in Era 206

<u>Sonar</u>

Sonar, short for Sound Navigation and Ranging, is defined as a system for the detection of objects under water and for measuring the water's depth by emitting sound pulses and detecting or measuring their return after being reflected. It is also noted to be the method of echolocation used in air or water by animals such as bats and whales. In 1906, American naval architect Lewis Nixon invented the first sonar-like listening device to detect icebergs. During World War I (1914-1918), a need to detect submarines increased interest in sonar. In 1915 French physicist Paul Langevin constructed the first sonar to detect submarines.

Sonar has many peacetime uses. It helps mapping the ocean floor as sound waves travel farther in the water than do radar and light waves. NOAA (National Oceanic and Atmospheric Administration) scientists use sonar to develop nautical charts, to locate underwater hazards, to search for objects such as shipwrecks, and to map the seafloor itself. The ability to map the ocean floor eventually led to the abandonment of old, incorrect theories, and the establishment of the unifying theory of Earth science: plate tectonics. This will be discussed in Era 207 (Economic Recovery, World War II, The Truman Doctrine).

Radio Telescope

People have gazed at the skies throughout our existence. But without special instruments the stars were just points of light. Our knowledge of celestial objects improved dramatically with the development of spectroscopy, as has been described in Era 201 (War of 1812, Louisiana Purchase, Lewis an Clark). Spectroscopic analysis enables astronomers to infer the elements in a star's atmosphere, and the elements and compounds in gas and dust clouds between stars and the Earth. Astronomy would take another huge leap forward with the development of the radio telescope. Radio waves can reach us through clouds of gas and dust that stop visible light.

Karl Jansky was born in 1905 and came from a family of physicists. He graduated from the University of Wisconsin with a degree in physics and joined Bell Laboratories in New Jersey in 1928. Bell Labs struggled with static affecting trans-Atlantic radio transmissions. They wanted improved radiotelephone service and asked Jansky to explore the origins of the static.

Jansky scanned the sky with a rotating antenna to receive radio signals. His radio wave receiver was operational in the autumn of 1930. After recording signals for several months Jansky discovered three types of static: a weak signal caused by distant thunder storms, a more powerful burst due to local thunder storms, and a third type that produced a steady hiss of unknown origin. Jansky originally believed the steady hiss to be radiation coming from the Sun. He continued to investigate it for over a year and eventually discovered that the signal was not coming from our Solar System, but instead from the direction of the constellation of Sagittarius - the center of Milky Way. (We now know that a supermassive black hole with a mass four million times that of the sun lies at the center of our Milky Way galaxy). Radio astronomy was born and would help usher in an age of immense discovery of the nature of the cosmos.

The Ford Model T

Prior to the automobile, local travel was typically by horse or horse and buggy. The Ford Model T (which had nicknames such as the Tin Lizzie and Flivver) was produced by Ford Motor Company from 1908 through 1927. As the first affordable automobile it opened travel to common middle-class Americans, thereby having a major effect on the American lifestyle.

Until this time, automobiles were expensive, unreliable, and scarce. The affordability of the Model T was due in major part to Henry Ford's having devised assembly line production instead of individual hand crafting. The Model T was more than transportation. It was a symbol of the rising middle class and of America's age of modernization. In 1999 there was a "Car of the Century" competition, and the Ford Model T was named the most influential car of the 20th century.

Ford followed the Model T with a new design and called it the Model A, because the new car was such a departure from its predecessor. Ford's Model T was extremely successful, as 16.5 million were sold. Amazingly, as of 2012 that figure put it in eighth place on the list of most sold cars of all time.

Medical Advances in Era 206

Treatment of Diabetes with Insulin

Before insulin, diabetes was a feared disease and a virtual death sentence. Doctors realized that sugar worsened the condition, and their only option was to put the patients on very strict diets with minimal sugar. This treatment may have bought patients a few extra years, but never saved them.

Nineteenth century researchers found damage in the pancreas of patients who had died of diabetes. In 1869, German medical student Paul Langerhans located clusters of cells in the pancreas whose function was unknown. Some of these cells were eventually shown to be insulin-producing cells. In honor of their discoverer, the cell clusters were later named the islets of Langerhans.

In 1889, German researchers Oskar Minkowski and Joseph von Mering studied with dogs and discovered that the pancreas must have at least two functions: to produce digestive juices, and to produce a substance that regulates the sugar glucose. If a substance could be isolated, the mystery of diabetes would be solved. Progress, however, would prove to be slow.

In October 1920 in Toronto, Canada, Surgeon Frederick Banting identified a secretion produced by the islets of Langerhans and wanted to test it. He needed a laboratory and prevailed on Professor John Macleod at the University of Toronto, a leading diabetes researcher. Though skeptical at first, Macleod gave Banting a minimal laboratory and dogs on whom to conduct tests. Banting and an assistant, medical student Charles Best, began experiments in 1921. Removing the pancreas from a dog resulted in increased blood sugar, thirst and weakness. The dog had developed the symptoms we now recognize as indicative of diabetes. They named their substance "isletin" and, when injected into the diabetic dog, it alleviated the symptoms. By

181

giving the diabetic dog a few injections per day, Banting and Best could keep it healthy and free of symptoms.

The new results convinced Macleod that they were onto something big. He gave them more funds and a better laboratory. He also suggested they should call their extract "insulin." The work proceeded rapidly. In 1921 biochemist Bertram Collip joined the team. Collip's task was to purify the insulin for testing on humans. But on whom should they test? Banting and Best began by injecting themselves with the extract. They felt weak and dizzy, but were not harmed.

In January 1922 in Toronto, 14-year-old Leonard Thompson was the first diabetic sufferer to receive insulin. Leonard was near death but, with treatment, rapidly regained his strength and appetite. The team expanded their testing to other volunteers, who reacted just as positively as Leonard had to the insulin extract. The world now had a treatment for a deadly disease.

What We Learned about the Universe in Era 206

<u>Galaxies and the Expanding Universe</u>

Most astronomers 100 years ago believed that the whole universe consisted of just one sea of stars, our own Milky Way. Observers had also noticed spiral patches of light among the stars, and they termed them "spiral nebulae." In the 1920s, Edwin Hubble was among the first to recognize that there is a universe of galaxies located beyond the boundaries of our Milky Way. He also showed that our universe of galaxies is expanding. This discovery took place in a series of steps.

It was Henrietta Swan Leavitt who started astronomy on the path to understanding the nature of spiral nebulae. Not all stars exhibit constant light and energy output. All stars can be rated by their luminosity (intrinsic brightness) and variable stars can be rated according to the length of time it takes for them to brighten and dim (period). Some of these stars are classified as "Cepheid variable stars." In 1908 Leavitt was studying the relationship between the period and luminosity for Cepheids in the Magellanic Clouds. She found that there is a direct relationship between the periods and luminosity of these stars.

During the 1920s, Edwin Hubble studied a patch of light known at the time as the Andromeda nebula. Some astronomers believed the Andromeda nebula was a nearby solar system in the process of forming. Hubble observed Cepheid variable stars in this nebula. Building on Henrietta Leavitt's work, Hubble knew that these stars' brightness changes depended on their true luminosity. The ratio of true luminosity to apparent brightness gives an estimate of distance. Hubble studied these stars and found that the nebula is extremely distant. Hubble had thus identified a separate galaxy. We know it today as the Andromeda Galaxy, the nearest large spiral galaxy beyond our Milky Way.

In the description of Era 201 (War of 1812, Louisiana Purchase, Lewis and Clark), we noted that the spectroscope had been invented and that it would eventually revolutionize astronomy. This was the moment! Hubble examined the spectra of several galaxies. He was confused by the absorption lines, as they did not look familiar. In a stroke of genius, he realized that the lines were familiar, but were shifted toward the red end of the spectrum. The frequency of the light waves coming from other galaxies had been attenuated, or stretched out. This same phenomenon occurs with sound waves. You have noticed this when a moving vehicle with a siren approaches you and then moves away from you. As the vehicle approaches, the pitch of its siren goes up, and as it recedes from you the pitch drops noticeably. Hubble reasoned that the light waves were stretched out because other galaxies are moving away from us at high speed. Hubble and his colleagues compared the distance estimates to galaxies with their red shifts. On March 15, 1929, Hubble published his observation that the farthest galaxies are moving away faster than the closest ones. This insight became known as Hubble's Law. It was the first recognition that the galaxies are moving away from each other. The conclusion was that our universe is expanding, and this realization has transformed our understanding of the Universe.

<u>Human Origins - Discovery of Australopithecus</u>

This era saw significant progress in our knowledge of human origins. Fossilized remains of a primitive protohuman skull were discovered in 1934 in a lime quarry at Taung, South Africa. Raymond Dart, an Australian-born South African physical anthropologist and paleontologist, studied the fossil and recognized its humanlike features. Dart's analysis of fossil hominins (members of the human lineage) led to significant insights into human evolution.

In 1924 Asia was believed to be the cradle of humankind. Dart's work supported Charles Darwin's prediction that fossils of human ancestors would be found in Africa. Dart made the Taung skull the "type specimen" of a new genus and species, Australopithecus africanus, or "southern ape of Africa." Dart's claim, if true, would support the principle that some characteristics can evolve in advance of others. That principle was not widely accepted at the time, so Dart's analysis was met with skepticism. Dart lived to see his theories corroborated by further discoveries of Australopithecus remains in South Africa in the late 1940s and by the subsequent discoveries of Mary and Louis Leakey. Africa is currently accepted by most anthropologists to be the site of humankind's earliest origins.

<u>Quantum Theory</u>

Quantum theory is the theoretical basis of modern physics that explains the nature and behavior of matter and energy on the atomic and subatomic level. The reason for the name "quantum theory" is that energy, and energy states, were found to exist not in a continuous form, but rather in tiny packets, or "quanta."

Seventeenth century physicist and mathematician Sir Isaac Newton gave us classical physics, and with it a picture of an orderly Universe. Classical physics is good approximation of reality in the gravity and velocities in which we live. It served well until the 20th Century and, for most purposes, it still does. Early in the twentieth century Albert Einstein's theories of special and general relativity shook the foundations of physics. Einstein showed us the equivalence of matter and energy. He showed us that the passage of time depends on velocity and on gravitational fields. And he showed us that space-time is a four-dimensional system and that what we experience as gravity is the behavior of objects in four-dimensional space-time. Thus far Einstein's physics are a good approximation of reality in the macro world.

But in the 1920s and 1930s, Niels Bohr, Erwin Schrödinger, Werner Heisenberg and others discovered that the "micro world" of atoms and subatomic particles is nothing like the "macro world" of objects familiar to us in our daily lives. The subatomic world is in fact full of chaos, and not the precise clockwork suggested by classical theory.

Niels Bohr was a prime mover in the development of quantum theory. In 1913 he found that an electron traveling around an atomic nucleus was restricted to certain well-defined quantities of energy - in a multiple of a unit called a quantum. Based on quantum theory, electrons can move only in prescribed orbits. When an electron jumps from one orbit to another with lower energy, a light photon is emitted. Bohr's theory explained why atoms emit light in fixed wavelengths.

In 1926, Austrian physicist Erwin Schrödinger took Bohr's model of the atom a step further. Schrödinger developed what is known as the quantum mechanical model of the atom. This model does not predict the location of an electron, but only the probability of the location of the electron. This model has been described as a nucleus surrounded by an electron cloud. The probability of finding the electron is greatest where the cloud is most dense.

Werner Heisenberg is another major theorist in quantum theory. He is best known for his uncertainty principle. This principle states that certain pairs of variables cannot both be known with precision. This is not due to the skill of the investigator, but is due to the nature of the system. If a particle is forced to be in a precise position, then the particle's speed cannot be precisely defined. If its speed is measured, then by virtue of that measurement the particle's position cannot be determined.

Quantum theory and quantum mechanics defy our logic. Niels Bohr once said, "Anyone who is not shocked by quantum theory has not understood it." Richard Feynman, winner of the 1965 Nobel Prize for Physics, said, "I think I can safely say that nobody understands quantum mechanics." Although it defies day-to-day logic, quantum mechanics works and, along with general relativity, is one of two theories underpinning modern physics. General relativity explains the world of the very big (space-time and gravity), while quantum theory explains the world of the very small (atoms and their constituents). Together, the two theories provide at least an approximate explanation of the Universe.

Quantum theory has a wide array of applications. It is a factor in lasers, CDs, DVDs, solar cells, fiber-optics, digital cameras, photocopiers, bar-code readers, LED lights, computer screens, transistors, semi-conductors, spectroscopy, MRI scanners, et cetera. There are those who estimate that, in the developed nations, over twenty-five percent of gross domestic product is directly based on quantum physics.

The Restless Earth in Era 206

<u>1912 Eruption of Novarupta</u>

The 1912 Eruption of Novarupta in Alaska was by far the largest eruption of the 20th century, three times larger by volume than the 1991 eruption of Mt. Pinatubo. It was thirty times larger than the 1980 eruption of Mt. St. Helens. Despite the size of the eruption, it had a surprisingly low impact on global climate, with some cooler temperatures in the northern hemisphere and some effect on air clarity. It did not have the major impact on climate that one might expect from an eruption that was so explosive and had such a large volume of material ejected from it.

The Mississippi River Flood of 1927

In late 1926 and early 1927, months of heavy rain caused the Mississippi River to rise to previously unseen levels. In mid-to-late April, a levee broke on the Illinois shore, and the levee at Mounds Landing in Mississippi gave way. The entire levee system along the river collapsed in the next few weeks. In some places, occupied areas were under 30 feet of water. The floodwater did not completely subside for more than two months. More than 23,000 square miles were submerged. About 250 people died and hundreds of thousands were displaced. Called either the Mississippi River flood of 1927 or the Great Flood of 1927, it was among the worst natural disasters in the country's history.

Miscellaneous Events in Era 206

Refrigeration for Home Use

During the nineteenth century consumers preserved their food in iceboxes with ice purchased from ice harvesters. These iceboxes were used until at least 1910 and there were no improvements. Consumers who used the icebox in 1910 had the same moldy, smelly situation that consumers had in the early 1800s.

General Electric (GE) was one of the first companies to overcome these challenges. In 1911 GE released a household refrigeration unit that was powered by gas. This eliminated the electric compressor motor and decreased the size of the refrigerator. However, GE's electric company customers were not happy, so GE developed an electric model. In 1927, GE released the first refrigerator that ran on electricity.

In 1930 Frigidaire, one of GE's main competitors, synthesized Freon. With the invention of synthetic refrigerants based mostly on a chlorofluorocarbon (CFC) chemical, safe refrigerators were possible for the home. Freon enabled the development of lighter, smaller, cheaper refrigerators. The average price of a refrigerator dropped from $275 to $154 with the synthesis of Freon. With this lower price, fifty percent of American households could afford refrigeration. Household life had thus taken a step toward resembling today's lifestyle.

Summary Comments on Era 206

(Years: 1909 – 1933)

As was true for Era 205 before it (Telegraph, Populist Movement, Pure Food and Drug Act), Era 206 was a time of great advances in medicine, technology, science and our understanding of the history of life on Earth and of the Universe.

Diabetes Mellitus is a pernicious disease that, when unchecked, can threaten a sufferer's heart, kidneys, and vision, and can lead to limb amputation and death. Insulin was developed as a remedy during this era. Sonar was developed and would lead to more than just wartime use to detect submarines. It also revolutionized Earth science by allowing us to map the ocean floors. The study of human origins was given a boost by the first discovery of the fossilized remains of an australopithecine. Australopithecus lived from about four million until two million years ago and is accepted as an early human ancestor. Through the research of Henrietta Leavitt and Edwin Hubble, our conception of the Universe was totally transformed. We now knew that our Milky Way galaxy is but one of countless billions of "island universes," or galaxies, and that the Universe is expanding. The discovery by Barnum Brown of Tyrannosaurus Rex skeletons gave us more knowledge about the history of life on the planet, and helped to capture the imagination of a new generation.

The quality of life for Americans was also transformed by invention and innovation. Henry Ford's development of the Model T and of assembly line production made automobile transportation available to millions. Home refrigeration revolutionized home food storage. There was an upsurge in the emphasis on entertainment, and "talkies" increased interest in movies. Most homes had radios, which ushered in a new wave of music and of shared culture, and baseball became the national pastime.

Era 206 saw a huge economic boom that was fueled in part by huge excesses of leveraged borrowing. This was followed by the Great Depression. Some of the country's finest and worst Presidents served in Era 206. The 27th President, William Howard Taft, was not a charismatic figure but, as did his predecessor, believed that the good of the people, and not private business, should be the business of government. He thus put aside lands rich in coal and oil for the benefit of all the people.

Woodrow Wilson was one of the most intelligent men to ascend to the presidency. His presidency was a major victory for the labor movement. Labor strikes had until then been illegal, and companies, states and the federal government had on many occasions used hired Pinkerton detectives, state militias and federal troops to violently break up strikes. With Wilson's encouragement, Congress passed legislation making strikes legal, giving workers bargaining power they had never possessed.

Wilson also helped farmers and workers by convincing Congress to lower tariffs. Though businesses could not charge as much for their goods, reduced prices helped the common people.

President Wilson attempted to keep the U.S. out of World War I, but war became inevitable when Germany attacked U.S. ships. Wilson is famous for having proposed a fourteen-point peace plan after the war. A centerpiece of the plan was the League of Nations, which did form. However, as Wilson had become ill and unable to promote the plan, Congress did not pass it. Without the United States, the League of Nations was destined to fail. Wilson's sponsorship of the Alien and Sedition Acts of 1917 and 1918, intended to ensure success in the war, would later have unintended and devastating effects.

The presidency of Warren Harding was President Grant revisited. Harding appointed old friends from Ohio to important posts, and they profited. Some referred to his administration as "The Ohio Gang." Harding died in office and scandals, such as the infamous Teapot Dome scandal, were uncovered after his death. Calvin Coolidge had won attention in 1919 when he approved of the National Guard breaking up a police strike in Boston. While serving as President he stated, "the business of America is business," and he allowed business free reign. This was an era of enormous excess in business speculation and leveraged borrowing. Despite warnings from others, including his successor, Coolidge did not recognize the dangerous excesses and took no action, thus helping lead to the Great Depression.

Coolidge was succeeded by Herbert Hoover, who had a record as a humanitarian, but is among the most misunderstood and unfairly maligned Presidents. He raised money to help travelers return home from abroad in World War I, personally donating much of the money for that cause. He helped bring food from the United States past German lines to save millions of Belgians from starvation. Eight years after he left office, he helped bring food to the Finns, who were at war with Russia. Nonetheless, the stock market crash of 1929 and the ensuing Great Depression were unfairly blamed on him as he was in office at the time. Although he clearly believed in and practiced humanitarian efforts as a private citizen, as President he was slow to abandon his conviction that government should not interfere with business or provide direct relief to people. This is seen as his greatest failing. It delayed his using government funds to start public works projects, and he lost the election of 1932.

Era 206 saw the rise of the Progressive movement. The Progressive era saw the passage of amendments sixteen through twenty to the Constitution. These amendments established the income tax, direct election of senators, alcohol prohibition and women's right to vote. Clearly, alcohol prohibition was unwise. However, in fairness to those who promoted it, it should be noted that alcohol consumption in the late nineteenth century was three times today's levels. Many a career and family undoubtedly did fall victim to alcoholism.

The Progressive movement was not an entirely consistent movement. But its main thrusts were the use of science and expertise in determining governmental policy, elimination of corruption, consumer protection, assistance to those victimized by industrialization, standards for professions such as medicine and law, and regulation of business to achieve power balance. Above all, the Progressive movement emphasized a view of government not as a problem, but rather as the means to solve problems.

ERA: 207

Years: 1933 – 1953

(Economic Recovery, World War II, The Truman Doctrine)

Presidents: Franklin Delano Roosevelt and Harry Truman

1932				
Franklin D. Roosevelt John Nance Garner	New York Texas	Democratic	472	57%
Herbert C. Hoover Charles Curtis	California Kansas	Republican	59	40%

1936				
Franklin D. Roosevelt John Nance Garner	New York Texas	Democratic	523	61%
Alfred M. Landon Frank Knox	Kansas Illinois	Republican	8	37%

1940				
Franklin D. Roosevelt Henry A. Wallace	New York Iowa	Democratic	449	55%
Wendell L. Willkie Charles McNary	Indiana Oregon	Republican	82	45%

192

1944				
Franklin D. Roosevelt Harry S. Truman	New York Missouri	Democratic	432	53%
Thomas E. Dewey John W. Bricker	New York Ohio	Republican	99	46%

Franklin Delano Roosevelt

Franklin Roosevelt lived from 1882 to 1945 and served as our 32nd president from 1933 to 1945. He is the only U.S. president to be granted by congress the right to serve more than two terms. Roosevelt is one of the best-known American Presidents. He helped the country emerge from the Great Depression and led the nation through World War II. He became a beloved figure throughout the world. His leadership in World War II prompted Congress to approve his running for a third and fourth term.

Roosevelt was brought up on an estate in upstate New York and in his family's second home in New York City. His was a wealthy family who traveled extensively in Europe. In his early years he was educated by tutors and learned foreign languages. By his mid-teens he was enrolled in a private school, the Groton School in Massachusetts, in which most students were from wealthy families. He attended Harvard University, where he continued to associate primarily with persons from elite families. His social consciousness would be developed later. After graduating from Harvard in 1903, Roosevelt entered Columbia Law School. He married Eleanor, who was destined to be a very important First Lady. In 1907 Roosevelt passed the bar and practiced law.

Roosevelt did not have early political ambition. He began his political career by running for the New York State Legislature at the request of the Democratic party. He ran in a solidly republican county and, as he ran as a democrat, his victory was noteworthy and a precursor to his subsequent popularity. Like his cousin, former President Teddy Roosevelt, Franklin believed that politics was, for a wealthy man, not an avenue to acquire more wealth, but rather an arena in which to serve his country.

In 1913 President Woodrow Wilson appointed Roosevelt assistant secretary of the navy, and he was gratified to serve in the same post his cousin Teddy had held. Roosevelt wanted to go to war in World War I as a fighting man, but President Wilson insisted he needed him in his administrative role. FDR, as he came to be known, feared what the world would be like with Germany under a Nazi dictator, and he supported the U.S. entering the war. He supervised an increase in the country's battleship fleet and devised plans for naval action and for a mine field blockade to keep German submarines in port.

FDR ran for Vice President on the ticket with Ohio Governor James Cox, but they were defeated. In 1921, at age thirty-nine, Roosevelt fell ill with polio, which would change his life. It took him three years of hard work, dedication and sheer willpower to become functional again. He could never again walk without braces and crutches, but this did not stop him from ascending to the presidency or from being a major player on the world stage. The public did not realize the extent of his physical handicap, as it was hidden by his aides and a cooperative press.

FDR was elected governor of New York in 1928. Soon thereafter the Great Depression began. Roosevelt used his power as governor to aid businesses and to aid people out of work. Radios were becoming common in American homes, and he talked to the people on the radio in what became known as the "fireside chats." His chats made him well-known and the Democratic Party made him their nominee for President in 1932. In the presidential campaign, FDR advocated unemployment relief, lower tariffs and the repeal of prohibition. He also wanted to curb the ability of powerful interests to exploit the people. Though President Hoover had not been responsible for the onset of the Depression, the people nonetheless blamed him, and Roosevelt was easily elected.

The years ahead were characterized primarily by the Great Depression, and by the efforts Roosevelt made to help the country emerge from it. Though the federal government had always been pro-business – sometimes subtly and sometimes by violently suppressing labor – there was no accepted method for stimulating the economy. The philosophy of keeping government hands off business and the economy was still a prevailing view, and Roosevelt was influenced by that view. Thus, he may not have gone far enough in his initiatives to get the country moving, and he prematurely removed the supports he did put in place when the economy appeared to be recovering. Nonetheless common folk loved Roosevelt as much as the established elite hated him.

Roosevelt prepared for his presidency by surrounding himself with experts in many fields, and the press referred to his advisors as his "brain trust." He appointed Frances Perkins secretary of labor - the first woman to a hold a cabinet post. Relief and recovery were the thrust of legislative acts passed during FDR's first one hundred days in office. FDR addressed the dual purposes of employment plus environmental protection with the passage of the Civilian Conservation Corps. This resulted in a quarter million young persons working in forests and parks to build roads, plant trees and work to control floods. States were helped to provide programs for unemployed persons through the Federal Emergency Relief Administration. The act establishing the Tennessee Valley Authority protected natural resources and saved that region from chronic poverty. And FDR strove to reduce government costs and balance the budget via the Economy Act.

Later initiatives included the National Industrial Recovery Act, the National Labor Relations Act, the Emergency Relief Appropriations Act and, significantly, the Social Security Act of 1935. Some of the legislative acts aimed to provide relief and recovery were later ruled unconstitutional by the Supreme Court. This resulted in a failed attempt by FDR to ask Congress to allow the President to appoint a new justice for every one attaining age seventy.

In a radio address during a banking crisis, Roosevelt made the famous statement, "The only thing we have to fear is fear itself." The President's confidence influenced the people, who had faith in his reforms and overwhelmingly supported him in the election of 1936. Roosevelt fervently believed in the government's responsibility to help "the forgotten man." He called his administration "The New Deal," a phrase to which people still refer (That tag was undoubtedly inspired by his cousin Teddy Roosevelt's "Square Deal"). Historians state that common people not only believed in Roosevelt, but felt they knew him.

There was, however, a faction of wealthy and influential Americans who believed that government intervention in the economy would destroy the country. They hated Roosevelt as much as small business owners and wage earners loved him. Some detractors would refer to a bad storm as "a Roosevelt." But events in Europe would soon take the nation's attention away from its domestic issues.

World War II began in 1939. Roosevelt hoped the allies would prevail and that the United States could avoid military involvement. Roosevelt sent supplies to Britain as other countries fell to Germany. In 1940 Roosevelt was nominated for what no President had served: a third term. There was a lot of anxiety about the war, but Roosevelt inspired confidence and he won the election easily. On December 7, 1941 the Japanese bombed Pearl Harbor and the United States was at war. Roosevelt and British Prime Minister Winston Churchill met often to plan strategy. The United States mobilized its armed forces, and by 1942 we were engaged in war actions in the Pacific, European and Mediterranean theatres. The war was still in progress for the election of 1944 and Roosevelt was elected yet again. The country did not want to trade in his leadership for another at this crucial time.

The long years of difficult service took their toll on Roosevelt's health. He used his remaining energy for one last, major project. In the summer and fall of 1944 representatives of the United States, Great Britain, the Soviet Union and Nationalist China met in Washington. D.C. They began planning the United Nations, whose charter was drawn up later, when Harry Truman was President. On April 12, 1945 Roosevelt died in his cottage in Georgia, and Truman succeeded him as President. People all over the world mourned the passing of Roosevelt, who may have been one of the most widely loved political figures in history. Twenty-six days after Roosevelt's death the Allies accepted Germany's surrender. Japan fought on, and surrendered only after the United States displayed the destructive power of the atomic bomb.

1948				
Harry S. Truman Allen W. Barkley	Missouri Kentucky	Democratic	303	49%
Thomas E. Dewey John W. Bricker	New York Ohio	Republican	189	45%
Strom Thurmond Fielding L. Wright	South Carolina Michigan	States Rights (Dixiecrat)	39	2%
Henry A. Wallace Glenn H. Taylor	Iowa Idaho	Progressive	0	2%

Harry S. Truman

Harry Truman lived from 1884 to 1972 and served as our 33rd president from 1945 to 1953. Until age seven Truman grew up on a farm near Harrisonville, where his outdoor activities were somewhat restricted by poor eyesight. His family then moved to Independence, Missouri, where young Harry could receive better schooling. He was an avid reader, and it has been said that by the time he was in his mid-teens he had read every book in the town's library. In 1900 Harry's father took him to the Democratic National Convention in Kansas City. He had the opportunity to run errands for delegates, and was enthralled by the highly charged atmosphere of the convention. Due to financial losses, Harry's father had to relocate the family from Independence to Kansas City just after Harry's high school graduation in 1901.

Harry wanted to attend college, but his family's finances precluded it. He worked for a railroad, as a bank teller, and on his family's farm. He joined a national guard unit, and during World War I he served in France as an artillery captain. He returned from the war and was among those feted in a triumphant parade. He married Elizabeth "Bess" Wallace, and the couple would have one child. Truman and a friend opened a men's clothing store that subsequently failed. Truman paid off his debts despite the availability of relief from debts through bankruptcy. His attitude of honesty and conscientiousness would later draw attention.

Truman went into politics and in 1927 was elected county judge. In that capacity he ensured that roads in his county were built honestly, and this drew further attention to him. In 1934 Truman was elected to the U.S. Senate. He was put in charge of a committee charged with overseeing how government funds were spent on supplies, and he is reported to have saved the country over a billion dollars. Truman supported President Roosevelt's New Deal. He made a speech disparaging greed and self-interest, and he blamed excessive concentration of wealth and power for unemployment. He won re-election to the Senate by a narrow margin in 1940. He fought against government waste and, through his efforts, the Senate created the Committee to Investigate the National Defense Program. The committee was informally known as "The Truman Committee." The committee saved billions of dollars of taxpayer money, and it did much more. It exposed poor quality steel and engines, and it recommended design changes that upgraded military equipment. An article in *Time Magazine* referred to The Truman Committee as "the watchdog, spotlight, conscience, and sparkplug to the economic war-behind-the-lines."

In 1944 the Democrats, uncomfortable with Vice President Henry Wallace's sympathetic stance toward Russia, asked President Roosevelt to drop Wallace from the ticket. in a compromise move, Truman was named by Roosevelt as his vice-presidential running mate. The ticket won the election and Truman became President upon Roosevelt's death on April 12, 1945. Earlier that year, from February 4 – 11, Roosevelt had met at Yalta near the Crimea with his World War II Allies, Britain's Winston Churchill and the Soviet Union's Joseph Stalin, to discuss the post-war reorganization of Germany and the rest of Europe. Immediately on assuming the presidency, Truman studied all the correspondence between Roosevelt, Churchill and Stalin. Having studied those materials and having consulted his own advisors, Truman noted that Stalin was breaking agreements he had made at the Yalta Conference. Specifically, Truman concluded that Stalin reneged on his agreement to allow free elections in Poland. On April 23, 1945 Truman met with a Soviet foreign minister. The meeting may have been confrontational, and Truman demanded that Stalin live up to his agreement. The United States had been expanding westward and the Soviet Union was expanding eastward, and some observers Had predicted the two growing superpowers would inevitably confront one another. The stage was set for a rivalry between the United States and the Soviet Union – a rivalry that would dominate world politics for seven decades.

The fighting in Europe had ended a month before Truman took office, and he went to Europe to help draw up peace plans. But the war with Japan was still raging, and Japan controlled China. The United States had developed the atomic bomb. If Truman did not use it, he would have had to order an invasion of Japan, which would have caused the loss of many American and Japanese lives. Use of the bomb would extinguish many Japanese lives. Truman was faced with what can aptly be described as one of the most terrible decisions any human being has ever had to make. On July 26, the American, Chinese and British governments issued the Potsdam Declaration. Japan was called upon to surrender or face "utter destruction." When unconditional surrender was not forthcoming, Truman ordered the use of the bomb, and the cities of Hiroshima and Nagasaki were destroyed before Japan surrendered.

Truman, along with his Secretary of State George Marshall, developed the European Recovery Act, which came to be known as The Marshall Plan. It was announced on June 5, 1947. The goals of the plan were to rebuild war-torn areas, remove barriers to trade, modernize industry, and improve European prosperity. The United Kingdom, France and West Germany were the major recipients of United States aid. Russia objected to what they saw as American interference in Europe and would not participate.

After the war Truman went to work on his domestic agenda. He tried to improve conditions domestically by promoting measures that would later be called the "Fair Deal." He favored health insurance for all Americans, an increase in the minimum wage and a guarantee of equal rights. However, Congress did not cooperate.

On March 12, 1947 President Truman unveiled the Truman Doctrine, which stated that America's policy would be to stop Soviet expansion. Stopping the spread of non-democratic governments became the overriding aim of U.S. foreign policy from 1947 through 2016. Russia, which had been a U.S. ally fighting against Germany in World War II, was starting to be aggressive and imperialistic. The stage was being set for events and policies that would dominate world politics into the present. The doctrine of communism was being espoused. Communist rebels were gaining power in Greece, and Truman feared that if Greece fell to communism, Turkey would soon follow. He convinced Congress to approve aid to Greece and Turkey, and the rebels were defeated. This author believes that although communism is at its heart a noble and egalitarian ideal, it has never existed, and perhaps never could exist, in our world. A short version of the definition of communism: "a society in which all property is publicly owned and each person works and is paid according to their abilities and needs."

In this author's opinion, countries known as "communist" are, in reality, totalitarian dictatorships and have never had any significant resemblance to the doctrine of Communism. Thus, I will write "communist" with a small "c" when referring to those countries.

New York governor Thomas Dewey ran against Truman in 1948, and Dewey was expected to win. But Truman campaigned diligently and won re-election. Truman's second term in office was dominated by the Cold War. Truman felt that if the U.S. were to provide aid to underdeveloped countries, those nations would not be susceptible to communism. He persuaded Congress to allocate money and technical assistance to vulnerable countries. In 1949 Truman provided military aid to western Europe, and was instrumental in creating NATO – the North Atlantic Treaty Organization.

Regardless of its true philosophy and methods, communist Russia sought to spread its domination, and U.S. foreign policy, beginning with the Truman Doctrine, has had the major goal of containment of communism. Korea was divided, with the North being communist and the South being supposedly democratic. When North Korean troops invaded the South in June of 1950 Truman sent in troops to defend the South. Thus began an undeclared war often referred to as "the Korean conflict." The conflict did not go particularly well, especially after American forces pushed north of the Yalu River, resulting in Chinese forces entering the fray and pushing the Americans back. Truman wanted to negotiate, but General Douglas MacArthur wanted to conquer northern Korea and use nuclear weapons against the

Chinese. MacArthur essentially wanted to make foreign policy. Truman fired MacArthur and later wrote, "If there is one basic element in our Constitution, it is civilian control of the military. Policies are to be made by elected political officials, not by generals or admirals."

Harry Truman left the White House in 1953 and lived another nineteen years. He and his wife Bess returned to Truman's hometown of Independence, Missouri, where he oversaw the completion of his presidential library. Upon the library's completion, Truman spent much time at his office there, until health concerns forced him to remain home. At the library, Truman enjoyed receiving important guests, meeting scholars studying his presidency, and speaking to groups of visiting school children.

Truman found time to relax and rest in his post-presidential years. He liked bourbon and enjoyed sharing it with friends, political allies, and dignitaries who came through Independence. While his health permitted, he took regular walks around town. Harry S. Truman died on December 26, 1972, of old age rather than any specific sickness. Bess vetoed plans for an elaborate state funeral and arranged an Episcopalian service in the auditorium of the Truman Library. Truman is remembered for his firmness in dealing with the Soviet Union, and for being an honest leader who took responsibility for his decisions. Most historians have asserted that Truman's decisions on awesome matters were correct ones.

Inventions that Changed our Lives in Era 207

The Transistor

A transistor is a semiconductor device used to amplify or switch electronic signals and electrical power. It is the key active component in practically all modern electronics. Many followers of the history of science and technology view it as one of the greatest inventions of the 20th century. It can be mass-produced using a highly automated process that achieves an astonishingly low per-unit cost, and that is a major reason for its ubiquity.

Before transistors were developed, vacuum tubes were the main active components in electronic equipment. The transistor has a multitude of advantages over vacuum tubes, including lower power consumption and cost, and greater safety, ruggedness, flexibility and efficiency. There are at least three dozen different types of transistors, which electronics engineers categorize into at least a dozen categories. Transistors typically have at least three terminals for connection to an external circuit. Transistors are made of semiconductor materials such as silicon, germanium or gallium arsenide. There materials allow a charge carrier – usually an electron – to carry a signal.

Transistors are switches, and switches do more than just turn lights on and off. Switches are grouped together into what are known as "logic gates," which are grouped into "logic blocks," which are grouped together into "logic functions," which are grouped together into chips. Did you follow that? (I didn't). Suffice it to say that multitudes of transistors are involved in the functioning of most of the tools we use to navigate our daily lives.

Medical Advances in Era 207

<u>Penicillin</u>

The introduction of penicillin in the 1940s began the era of antibiotics, viewed as one of the greatest advances in therapeutic medicine. Before penicillin there was no effective treatment for strep throat, bronchitis, pneumonia, rheumatic fever, or for a host of other infections. Hospitals were full of people with infections from a cut or scratch, and doctors could do little but wait and hope.

Penicillin was discovered in London in September of 1928. The story is that Dr. Alexander Fleming, the bacteriologist on duty at St. Mary's Hospital, found a messy lab bench on his return from a summer vacation. While sorting through petri dishes containing colonies of Staphylococcus, he noticed one dish dotted with colonies, except for an area where a blob of mold was growing. The area around the mold was clear, as if the mold had inhibited the bacteria. Fleming found that this mold could kill many harmful bacteria, such as streptococcus, meningococcus and the diphtheria bacillus.

At first penicillin was viewed primarily as a research tool. Eleven years after Fleming's discovery, researchers at Oxford University developed methods for growing greater amounts of penicillin, and then experimented with its therapeutic use with mice. During World War II, efforts were made in both Britain and the U.S. to develop the treatment for use on the battlefield. It was a long road for researchers and pharmaceutical companies to refine their methods so as to produce enough of the substance to make it generally available. Finally, by March 15, 1945, penicillin was distributed through usual channels and was available to physicians and to consumers in pharmacies. This was the beginning of the age of antibiotics. As will be discussed in Era 210 ("9/11" Twin Towers attack, Second Persian Gulf War, Climate Change), we now know that we are engaged in a war with various bacteria that can mutate, perhaps faster than we can develop new antibiotic weapons to fight them. Nonetheless, for seventy years illnesses have been alleviated and lives saved for millions of persons using antibiotics, which began in 1945 with penicillin.

The Restless Earth in Era 207

<u>The Dust Bowl</u>

As stated in the introduction, events under the heading "the restless Earth" include geological events as well as impacts from space and occurrences of human origin. One example of the latter type was a devastating series of dust storms that took place in the U.S. and Canadian prairies in the 1930s and into 1940. The dust storms took place in three waves – 1934, 1936 and 1939-40 – and they were eventually termed "The Dust Bowl." A recent documentary describing history's ten worst weather-related disasters ranked the Dust Bowl as the most devastating of all.

The devastating dust storms were caused by a lack of advanced agricultural methods. For at least a decade, farmers used deep plowing of the virgin topsoil, displacing deep-rooted grasses that had covered and protected soil and moisture during periods of reduced rainfall.

The unprotected soil turned into dust that was at the mercy of winds. It has been reported that huge clouds of dust blackened the sky. These events were termed "black blizzards" or "black rollers" that sometimes travelled as far as New York City and Washington. They not only devastated agriculture, but were an enormous health hazard. People could often not leave their homes without trying to cover their heads and faces, often to little avail. Dust choked cattle, and lung diseases were prevalent. Many families abandoned their farms and migrated. It is estimated that sixty percent of the population from the affected area departed.

In his 1939 novel *The Grapes of Wrath*, John Steinbeck wrote, "And the dispossessed were drawn west – from Kansas, Oklahoma, Texas, New Mexico; from Nevada and Arkansas, families, tribes, dusted out. Car-loads, caravans, homeless and hungry; twenty thousand and fifty thousand and a hundred thousand and two hundred thousand. They streamed over the mountains, hungry and restless – restless as ants, scurrying to find work to do – to lift, to push, to pull, to pick, to cut – anything, any burden to bear, for food. The kids are hungry. We got no place to live. Like ants scurrying for work, for food, and most of all for land."

The Dust Bowl caused hardship, death, disease and the relocation of tens of thousands of families. Perhaps the farmers of that era could not have been expected to foresee the consequences of their methods. Nonetheless, it is an example of an axiom in the biological sciences: "You can never do one thing." In our current era, with approximately 7.9 billion people on the planet, this truth has still not impressed itself on enough citizens and policy-makers.

What We Learned about the Universe in Era 207

Earth Science – The Earth's Core

For centuries the inner structure of our Earth was a mystery. Over the course of two centuries the mystery has gradually been uncloaked. In 1798 Lord Cavendish used measurements of gravitational attraction to calculate the Earth's density. His result - 5.48 times the density of water – is remarkably close to the modern measured value of 5.53. One hundred years later, Emil Wiechert reasoned that Earth cannot be composed entirely of rock as the density of the Earth is greater than the density of rock. Some large meteorites have nickel-iron cores, and in 1898 Wiechert suggested that the Earth might be similar. He calculated that the Earth's density could be explained if Earth has a core of nickel-iron metal surrounded by a mantle of rock.

In 1906 English seismologist Richard Oldham studied the speed of earthquake waves at different depths in the Earth. He found that the greater the depth, the faster the waves. But below a certain depth, earthquake waves slowed down. This suggested a different material at great depth. Oldham had confirmed that Earth has a core, as Emil Wiechert had surmised. In the 1930s, seismologists estimated the size of the Earth's core, and it appeared to be molten.

But another mystery came to light. Earthquake waves were found where they should not be found. Danish seismologist Inge Lehmann solved that mystery. She analyzed the measurements and concluded that to match the measurements, Earth must have a solid inner core and a molten outer core. That is our current understanding. Earth has a solid inner core, with a molten outer core that is in constant motion. Significantly, the motion of the outer core provides Earth with its magnetic field, without which Earth would be devastated by the constant outflow of energetic particles from the Sun.

<u>Nuclear Fission</u>

Nuclear fission is the process by which an atom splits into lighter atoms, releasing considerable energy. Its discovery has had a profound effect on our world in the areas of energy, science, medicine, and world politics.

75 years ago, three scientists, Drs. Otto Hahn, Lise Meitner and Fritz Strassman, working in Berlin, experimented with the new concept that splitting an atom would produce two atoms of smaller, different elements. Compared to today's experimental laboratories, their work was elementary, but it was cutting-edge research at the time. A neutron source was sealed in brass tubes and placed in a paraffin block, slowing the neutrons. They used the neutrons to irradiate a uranium sample. As the neutrons bombarded the uranium sample, nuclear fission occurred. The decay of the radioactive substances was measured with home-made Geiger-Müller radioactivity counters.

Meitner's Jewish ancestry made it unsafe for her to stay in Berlin due to the political situation in Germany. In 1938 she was obliged to relocate to Sweden. Hahn and Strassman continued research and shared their results with Meitner. She and her nephew, physicist Otto Frisch, correctly interpreted the data and explained how uranium nuclei split into lighter elements, releasing neutrons and large amounts of energy. The threesome coined the process "nuclear fission" as it was comparable to the biological fission observed in cell division.

Hahn, Meitner and Strassman declared fierce opposition to nuclear technology for military use. They refused to be involved in the development of the weapons their discovery made possible. The discoverers of nuclear fission asserted that their findings' great potential lay in their peaceful applications.

Miscellaneous Events in Era 207

The Great Depression and The New Deals

When Franklin D. Roosevelt ascended to the presidency in 1932 the U.S. was in a deep economic depression. The stock market had soared upward, propelled by unbridled optimism and unwise leveraging of assets. But it had crashed in October of 1929. Banks failed, millions of people were out of work, and the economy stagnated. Roosevelt's first goal was to revive the economy, and his first intervention was psychological. He succeeded in his effort to restore the public's confidence and to stop runs on the banks. He declared a "bank holiday" after his inauguration and, on March 12, 1933, he held the first of his famous radio-broadcast "fireside chats." He told the public the banking system was sound and banks would re-open the next day. People lined up at the banks and made deposits, and the immediate financial/psychological crisis was over.

Roosevelt's major goals were the "three Rs" of recovery, relief and reform. Early in his administration he proposed fifteen bills to Congress, and all were passed in what has been termed "The First New Deal." They included the Homeowner's Loan Act, the National Industrial Recovery Act and the Agricultural Adjustment Act. These measures stopped the economy's downward spiral, but the recovery stalled by 1934.

Contrary to common myth, the First New Deal did not increase deficit spending, as Roosevelt and his advisors were still fiscally conservative. The First New Deal fell short of expectations for employment. And in 1935 the Supreme Court ruled that the National Industrial Recovery Act and the Agricultural Adjustment

Act were unconstitutional. Roosevelt and his advisors went back to the drawing board and implemented new measures

In the years that followed, in what has been termed "The Second New Deal," Roosevelt and Congress passed the Social Security Act, established the National Labor Relations Board and put millions of people to work on public works projects. It is estimated that 8.5 million people in total were employed on public works in the years 1935 – 1943. Roosevelt and his advisors all had roots in the Progressive era in that they believed the government does have the responsibility to solve problems and to help provide at least some security for all citizens.

The Second New Deal did briefly stimulate the economy and reduce unemployment. But Roosevelt incorrectly assumed the recovery was complete, and he cut government spending and attempted to balance the federal budget. The result was another major recession in 1937-1938. Full recovery would not be achieved until 1941, stimulated by World War II.

<u>World War II</u>

No attempt is made here to give a thorough description or analysis of World War II. A few key points will be advanced. First, there is a common myth that despotic dictators in the so-called Axis alliance – Germany, Italy and Japan – seized power in their countries, banded together and waged war on the free world. Those three countries were never actually in any firm alliance, though they conspired in a diplomatic bluff to try to frighten their enemies into inaction. Moreover, Fascism was an appealing idea to many in that era. Its ideology was that true freedom lay not in individuality, but rather in acting as a group, with a single powerful leader to focus the will and actions of the group. Further, the three Axis leaders – Hitler in Germany, Mussolini in Italy and Hidecki in Japan – came to power with the acquiescence of the political structures in their respective countries.

Several unconnected regional wars raged in Europe and Asia from 1931 to 1941 and culminated in the world war. Axis total victory in 1940-1941 was stopped in part by American material support, but primarily by Chinese military action in Asia and British and Russian resistance in Europe. The United States entered the war militarily after the bombing of Pearl Harbor by the Japanese on December 7, 1941. Japan had achieved victories in the Pacific, and they bombed Pearl Harbor to try to remove the threat the American navy posed to their further advance.

The United States did mobilize fifteen million in uniform for the war, amounting to about twenty-five percent of allied forces. U.S. Military deaths totaled about 416,800 and U.S. total civilian and military deaths totaled 418,500. Russian military deaths were estimated to be between 8,800,000 and 10,700,000, and total Russian civilian and military deaths totaled about 24,000,000. Russian sacrifices, and contributions to the defeat of Germany, are typically underestimated.

The Marshall Plan and the Origin of the Cold War

World War II officially ended on September 2, 1945. The war had taken a tremendous toll on many European countries, economically and in terms of loss of lives and physical property. Loss of lives among Russians was particularly massive. The U.S. Secretary of State devised the European Recovery Program, also known as The Marshall Plan. The U.S. channeled over thirteen billion dollars to finance the economic recovery of Europe between 1948 and 1951. The stated purpose of the plan was "restoring the confidence of the European people in the economic future of their own countries and of Europe as a whole." The Marshall Plan did successfully spark economic recovery, thereby meeting its objective.

Americans perceived the plan as a generous measure for Europe. However, the Soviet Union viewed it otherwise. Many historians believed that, even while allies fighting against the Axis powers, the U.S. and the Soviet Union were natural enemies. The two countries had different histories and ideologies. In addition, each had been expanding, the U.S. toward the west and the Soviets toward the east. It was seen as inevitable by some observers that their expansionist activities would inevitably bring them into conflict. The Soviets sensed the emerging conflict, saw the Marshall Plan as interference in the affairs of other states, and would not participate. Although Poland and Czechoslovakia were eager to participate, the Soviets prevented them from doing so.

Some revisionist historians question the idea that the Marshall Plan displayed American altruism. They see it as an expression of self-interest, arguing that sending dollars to Europe provided a market for U.S. exports and stopped the United States from sliding into economic depression.

After World War II, there were two superpowers: the U.S. and the Soviet Union. The United States feared the spread of communism. The United Kingdom had been fighting communist insurgencies in Greece and Turkey. But when they notified the U.S. that they no longer had the resources to do so, U.S. President Harry S. Truman issued what would be known as the Truman Doctrine: a promise that the United States would do whatever was necessary economically and militarily to contain the spread of communism around the world. *Containment became, and remained until 2016, the driving doctrine of U.S. foreign policy.*

The Cold War

The Cold War, which is typically considered to have lasted forty-five years, was an open but militarily limited rivalry after World War II between the United States and the Soviet Union and their allies. Neither country believed it could militarily defeat the other. The Cold War was waged on political, economic, and propaganda fronts with sporadic and restricted use of weapons. Military conflicts were usually between surrogates and not by the United States and the Russians themselves. Though not used, nuclear weapons were nonetheless manufactured and stockpiled. By 1949 the Soviets had tested their first nuclear bomb, and the world has lived with the knowledge that weapons existed that could devastate vast populations of people in a virtual instant.

For a variety of reasons, the Soviet system eventually went into decline. Premier Mikhail Gorbachev saw this and called it a period of stagnation. He made comments about how the Industrial wheels were turning but the production was not there. Growth slowed down markedly in the early 1980s, their economy stagnated, and the country could not feed itself. During 1989 and 1990, the Berlin Wall came down, borders opened, and free elections ousted communist regimes in eastern Europe. In late 1991 the Soviet Union dissolved into its component republics. With surprising speed, the Cold War came to an end. But that may not have been the final chapter.

McCarthyism

The McCarthy Era is one of the most disruptive and dangerous events in our history. There was a strong thrust in the United States to avoid and combat communism, and this effort reached dangerous heights in this era. Communism is the philosophy, derived from the ideas of Karl Marx, that all property should be publicly owned and that each person should work and be paid according to their abilities and needs. Communism may be seen as a noble ideal. However, it appears to be inconsistent with a system encouraging incentive and innovation. And, it may well be unattainable given human nature. It should also be noted that, although some countries are and have been self-styled communist, they are not. They are totalitarian regimes and bear no resemblance to the ideals of communism.

What would later be known as the McCarthy era began before Joseph McCarthy became prominent. In 1938, the House Un-American Activities Committee (HUAC) was established. Its mission was to investigate alleged disloyalty and subversive activities on the part of private citizens. After World War II, the atmosphere of fear of the Soviet Union caused the HUAC to command broad popular support.

In 1947, President Truman ordered federal agencies to screen employees for ties to organizations deemed "totalitarian, Fascist, Communist, or subversive," or seeking to "alter the form of Government of the United States by unconstitutional means." The term "McCarthyism" was used for the first time in 1950 after Senator McCarthy presented a list of alleged members of the Communist Party working in the State Department. The HUAC was a committee in the House of Representatives. Though associated with the HUAC, McCarthy was actually the chairman of the Senate Government Operations Committee, which carried on similar investigations. What followed was an era of suspicion and paranoia, during which private citizens were called upon to spy on and offer testimony about the activities, affiliations and beliefs of their fellow citizens.

In this era, government committees abused power and trampled important First Amendment rights, such as freedom of expression and freedom of association. Perhaps even more important is what this era meant about the process of the development and discussion of ideas. If you have learned anything from this book, gentle reader, it is that despite its successes, the American economic, social and political scene has typically left many persons impoverished and underprivileged while others have profited immensely. It is only natural that people would think about ways to develop a more fair and equitable system. McCarthyism essentially prohibited people from openly exploring many such ideas.

By instilling suspicion and fear about the word "communism," McCarthyism effectively told people what to think, what to say, and with whom to associate. It may impossible to overstate how dangerous, divisive and nefarious these activities were. This era created an immense tear in the fabric of American society. The term McCarthyism has come to mean reckless, unsubstantiated accusations, and demagogic attacks, on the character or patriotism of political adversaries.

Summary Comments on Era 207

Years: 1933 - 1953

Era 207 was an era of great change, and of disruption of world affairs. It was dominated first by recovery from the Great Depression, and then by the presidency of Franklin Delano Roosevelt, World War II, and by the beginning of the rivalry between the United States and Russia. The era was also dominated by the discovery of nuclear fission. This discovery led to the weapon that enabled the U.S. to defeat Japan to end the war, and to the stockpiling of weapons by the U.S. and the Soviet Union that partially dominated forty-five years of Cold War.

When FDR took office, the country was mired in the Great Depression. FDR's cousin and former President Theodore Roosevelt had established that the government could take an active role in solving the nation's problems. Roosevelt's first response to the Depression was to reassure the nation that the banking system was sound. He then initiated several legislative measures to get the economy moving. These measures included public works projects, many of which are still in evidence today.

Although Roosevelt attempted to keep the U.S. out of World War II, it became impossible when the Japanese attacked Pearl Harbor on December 7, 1941, which FDR called "a date that will live in infamy." The United States joined Great Britain, France and Russia and eventually defeated the loose association called the "Axis Powers": Germany, Italy and Japan. General Dwight Eisenhower commanded the allied forces as they landed in Normandy, France. Eisenhower's prominence during the war led to his later becoming the nation's thirty-fourth President.

Harry Truman became President upon Roosevelt's death. He and his Secretary of State George Marshall advanced the Marshall

Plan to aid in Europe's recovery. Russia was suspicious of the United States' motives and did not participate, and the seeds of the Cold War were sewn. Late in this era, the fear of an idea inflicted tears in the fabric of American life. Senator Joseph McCarthy exploited fears of communism and ushered in an era of suspicion, paranoia and injustice.

Advances continued to be made in Era 207 in technology, medicine and our understanding of the Universe. The invention of the transistor is considered by many to have been the premier invention of the twentieth century. It lies at the heart of most of our electronic and computer technology. In an earlier era, the germ theory of disease had been recognized. But it was in Era 207 that the breakthrough discovery of penicillin resulted in the alleviation of much suffering and death through disease.

Our understanding of our own Earth continued to develop during this era. Though it cannot be directly observed, the layers of the Earth, including its iron-nickel core, were beginning to be understood. The secrets of atomic nuclei, including radioactive decay and fission, were revealed, with devastating results at the end of World War II. Poor farming methods led to what has been rated the most devastating weather-related event in the nation's history: The Dust Bowl disasters between 1934 and 1940.

ERA: 208

Years: 1953 – 1974

(The Cold War, The Vietnam War, The Moon Landing)

Presidents: Dwight Eisenhower, John F. Kennedy, Lyndon B. Johnson and Richard Nixon

1952				
Dwight D. Eisenhower Richard M. Nixon	Kansas California	Republican	442	55%
Adlai E. Stevenson John J. Sparkman	Illinois Alabama	Democratic	89	44%

1956				
Dwight D. Eisenhower Richard M. Nixon	Kansas California	Republican	457	57%
Adlai E. Stevenson Estes Kefauver	Illinois Tennessee	Democratic	73	42%

Dwight David Eisenhower

Dwight D. Eisenhower lived from 1890 to 1969 and served as our 34th president from 1953 to 1961. He was born in Texas and grew up in Kansas. In school, Eisenhower liked spelling and math, and played football and baseball. But he was fascinated by military history, and would ignore schoolwork and chores in order to read about it. He had his sights set on a military career from an early age. His military service academy exam score fell short of requirements for the Naval Academy, but did ger him into West Point. He attended West Point and, though his grades were

216

average, he was well-liked.   His nickname – Ike – stuck with him and became part of the campaign slogan "I like Ike" later years.

Eisenhower graduated from West Point in 1915 and was assigned to Fort Sam Houston in Texas as a second lieutenant. He wanted to fight in World War I, but the war ended before Eisenhower was scheduled to go to Europe.   He served at several posts between the two world wars. He won an appointment to attend a military command college in Leavenworth Kansas and, though he had graduated in the middle of his class at West Point, he now excelled and graduated first in a class of 245.   His career accelerated when he served as military aid to Generals John Pershing and Douglas MacArthur.

In 1941 Eisenhower came to the attention of General George Marshall when he won an impressive victory in a simulated training battle. Marshall called Eisenhower to Washington to work as a planning officer after Japan's attack on Pearl Harbor. Now a Brigadier General, Eisenhower led successful invasions of North Africa, Sicily and the Italian mainland.   In 1943 General Marshall appointed Eisenhower commander of all the American forces in Europe.   He commanded Allied forces during the Normandy Invasion of June 6, 1944, an invasion that led to the liberation of Paris and that helped change the course of the war.

After the war Eisenhower resigned from active military duty and became president of Columbia University.   The country has a tradition of loving its military heroes, and both parties wanted him to run for President in 1948.   He stated that he did not believe that a military man should be President, and stated, "I haven't a political ambition in the world."   But Truman's popularity was waning, partly due to the Korean conflict, and the Republicans persuaded him to run for President in 1952. He later reflected on 1952, recalling that though he would advance the reasons he should not run for the presidency, his associates held a conviction that running was his duty. Eventually, he declared himself a Republican and made himself a candidate. For his running mate he chose Richard Nixon, then a California Senator who was a strong anti-communist. This played well to the country's anti-communist sentiment. Many voters wore buttons that said "I like Ike," and Eisenhower won the presidency.

Eisenhower was not a gifted speaker, but history regards him as having been honest, decent and as having had good intentions. But the international and domestic situations were complex and difficult. The French were involved in military activity in what was then called Indochina. Eisenhower's decision not to rescue French forces there may have had the unintended consequence of setting the stage for our long and ultimately unsuccessful involvement in the Vietnam War.

Eisenhower tried to complete the desegregation of the armed forces begun by Roosevelt and Truman. He went to Korea and helped end the conflict there. He attempted to work with the Soviet Union to restrict nuclear arms testing, but was unable to accomplish much in that domain. He asked Congress for new civil rights laws, but was unable to persuade Congress to enact them.

The nation had a fear of communism, and Senator Joseph McCarthy fanned the flames into dangerous paranoia. McCarthy held hearings intended to expose and discredit supposed communists and communist sympathizers. Citizens became wary of one another and, in some industries, such as the movie industry, many persons were "blackballed" and deprived of their livelihoods due to the activities of McCarthy's Government Operations Committee. Eisenhower disliked McCarthy's tactics and is said to have attempted to mitigate this problem behind the scenes, but he was ineffectual in this regard. His failure to openly oppose and suppress McCarthy's efforts may have been his greatest failing. The rampant paranoia spawned by McCarthyism nearly tore the fabric of the nation apart.

One of Eisenhower's major aims was to reach an agreement with the Soviet Union. But he allowed espionage efforts that would prove to be an embarrassment. He approved flights over the Soviet Union by high-altitude U2 spy planes even while he was arranging a conference on disarmament with Britain, France and the Soviet Union. Two weeks prior to the conference, a U2 piloted by Captain Gary Powers was shot down well within Soviet territory. The CIA assured Eisenhower that the plane would have been destroyed and Powers would have been killed, so Eisenhower publicly denied the spy plane's flight. Though humiliated by the lie, Eisenhower attended the conference. But Soviet premier Nikita Khrushchev lambasted Eisenhower and stormed out, ending hopes for a slowing of the arms race.

Another major development during Eisenhower's presidency was the beginning of "the space race." On October 4, 1957 the Soviets launched a satellite, known as "Sputnik." This shattered the American belief in U.S. technological superiority over the Soviets. In his farewell address of January 1961, Eisenhower stated that a partnership of the military and large business interests could exert too strong an influence on the course of American government. He warned us about the inherent dangers of what he termed the "military industrial complex."

Eisenhower retired to his home near Gettysburg, Pennsylvania, on the edge of the Civil War battlefield. He raised Angus cattle, painted, spent time with his wife Mamie and wrote his memoirs. The former President and his wife also traveled, and revisited the sites of past triumphs. They visited Normandy for the filming a documentary for the twentieth anniversary of D-Day. Though out of office, Eisenhower could not ignore national affairs. He occasionally provided advice to President John F. Kennedy, and he consulted frequently with President Lyndon B. Johnson after LBJ committed combat troops to Vietnam. He advised a strong attempt for victory in the Vietnam War - advice that later events would show to have been misguided. Eisenhower wrote his memoirs and a chatty best-seller, *At Ease: Stories I Tell to Friends*.

Eisenhower suffered a heart attack in 1965, and his health deteriorated. He was in Walter Reed Army Hospital for nine months until his death at age 78 on March 28, 1969. He went out like a general, commanding doctors and nurses to lower the shades and pull him up to a sitting position in bed before he died. People around the world mourned his death.

1960				
John F. Kennedy Lyndon B. Johnson	Massachusetts Texas	Democratic	303	49.7%
Richard M. Nixon Henry Cabot Lodge	California Massachusetts	Republican	219	49.5%
Harry F. Byrd	Virginia	Independent	15	

John Fitzgerald Kennedy

John F. Kennedy lived from 1917 to 1963 and served as our 35th president from 1961 until his assassination in 1963. John Fitzgerald Kennedy, called "Jack" by his family but typically known by his initials, JKF, was born in Brookline, Massachusetts to Joseph and Rose Kennedy. Joseph Kennedy rose from somewhat modest beginnings to great wealth. He profited from stock investments, bootlegging and an interest in a movie company. He was clever enough to know that the rising stock market in the late 1920s was a bubble, and he profited greatly from short sales. Though profiting from short sales was perfectly ethical, he was nonetheless known to engage in some unethical practices.

As JFK's father was absent due to his business priorities and his mother, Rose Kennedy, had a distant character, JFK grew up without parental warmth. He was treated at a sanitarium at age three due to scarlet fever, and battled several illnesses later. He attended Choate, an exclusive private school, where he played football and hockey. He attended Princeton briefly, then transferred to Harvard. His father had been appointed ambassador to Britain by President Roosevelt, and Jack visited

London and toured Europe in 1938. He returned to school with greater interest in academics, and he graduated cum laude from Harvard University in 1940 with a bachelor of arts in government. For his honors thesis Kennedy wrote a book entitled *Why England Slept*, in which he criticized Britain's policy of appeasement of Hitler. He wrote that if a democratic country were to be isolationist, it would threaten the continuation of democracy, and his book became a best seller.

Kennedy joined the U.S. Navy after the Japanese attack on Pearl Harbor. During World War II he commanded a series of PT boats and won Navy and Marine medals for his service. On one occasion in the darkness of night, his PT boat was rammed by a Japanese warship. Kennedy and several of his men persevered and made it to the shore of an island. To save another seaman, Kennedy is said to have towed the man to shore by gripping the man's life jacket strap in bis teeth.

After the war Jack's father encouraged him to enter politics. The elder Kennedy strongly supported his son's 1946 campaign for Congress, and distributed reprints of a magazine article describing Jack's heroism as captain of PT109. Jack, who by then was known as JFK, campaigned hard and won a seat in Congress. He supported the Truman Doctrine and the Marshall Plan, and early on he sympathized with the efforts of Senator Joseph McCarthy. He later broke with McCarthy when the latter's treachery became evident.

JFK won a seat in the U.S. Senate in 1952. In 1954 he had back surgery, and the surgery plus his Addison's Disease could have resulted in his death. From then on, he was never free of pain. Kennedy served seven years in the Senate. While in the Senate, and while recovering from a second back surgery, he wrote a book entitled *Profiles in Courage,* in which he lauded senators who had put principles before politics. The book earned JFK a Pulitzer Prize. In 1960 Kennedy won the Democratic nomination for President over Hubert Humphrey, and he chose Texas Senator Lynden Johnson to be his running mate. Kennedy and his

Republican opponent, Richard Nixon, engaged in the first ever televised presidential debates. The campaign was not without participation of machine politics in such places as Chicago. Kennedy defeated Nixon in an extremely close election and became the first Roman Catholic President as well as the youngest person to ascend to the presidency.

As President, Kennedy chose the most highly intelligent and qualified individuals for important posts. They included teachers and scientists, and many worked for less money than they had made before. They did so because they believed in Kennedy and in his vision for a better America. These appointees have been referred to as "the best and the brightest." Unfortunately, not all prior or subsequent Presidents have followed such a course.

Kennedy inherited a plan conceived under Eisenhower. The plan, developed by the CIA, was to aid Cuban refugees in their attempt to overthrow Cuba's Fidel Castro. In April of 1961, CIA-trained refugee battalions landed at Cuba's Bay of Pigs, and were crushed by Castro's forces. Kennedy had asked many advisors about the plan and received their approval. But the operation failed and damaged Kenedy's credibility.

As has been previously stated, the policy of containment started by President Truman and his Secretary of State George Marshall has dominated American foreign policy. The Cold War between the U.S. and the Soviet Union was a dominant theme during Kennedy's presidency. In October of 1962 American spy planes discovered Soviet missiles in Cuba. This precipitated a confrontation known as the "Cuban Missile Crisis." Kennedy confronted the Soviets with a show of force, and this may have been the closest the country has ever come to nuclear war. Ultimately, the Soviets backed down and removed the missiles they had installed. Some historians have called Kennedy's handling of the Cuban missile crisis his finest moment.

Kennedy's administration achieved a great deal on the domestic front, passing Important legislation in many areas: the economy, taxation, labor, education, health, civil rights, equal rights for women, the environment, agriculture, crime and defense. A few examples among many will be noted. Justice: 288 organized crime figures were brought to trial by a team headed by Kennedy's brother, Attorney General Robert F. Kennedy. Agriculture: The Rural Renewal Program enabled the USDA to provide technical and financial assistance to farming communities. Environment: The Clean Air Act was passed and programs to control water pollution were doubled. Employment: vocational educational programs were modernized. Housing: The Senior Citizens Housing Act made loans available for projects providing housing for persons over age sixty-two. Civil rights: government contractors were forbidden in employment matters to discriminate based on race, creed, color or national origin. The economy: the Trade Expansion Act authorized the President to reduce tariffs on a reciprocal basis with European countries.

When Kennedy was elected, he used the term "New Frontier," and in his acceptance speech he stated,

> We stand today on the edge of a New Frontier — the frontier of the 1960s, the frontier of unknown opportunities and perils, the frontier of unfilled hopes and unfilled threats. Beyond that frontier are uncharted areas of science and space, unsolved problems of peace and war, unconquered problems of ignorance and prejudice, unanswered questions of poverty and surplus.

Historian Robert D. Marcus wrote, "Kennedy entered office with ambitions to eradicate poverty and to raise America's eyes to the stars through the space program." Though Kennedy would not live to accomplish all his goals, his promise that we would attain the goal of putting a man on the Moon was indeed realized in 1969. Without his stirring encouragement, we might never have mobilized the resources to accomplish this milestone. On November 22, 1963, Kennedy was assassinated in Dallas, Texas by Lee Harvey Oswald. A government commission headed by

Supreme Court Chief Justice Earl Warren concluded that Oswald had acted alone, though competing theories circulate to this day.

JFK's widow, Jacqueline Kennedy, recalled that her husband favored music from the Broadway play *Camelot*. She called his presidency "A magic moment in American history," famously stating, "There will never be another Camelot." There was much admire about Kennedy and his presidency. He was inspirational, espoused high ideals, recruited dedicated, talented persons for his administration, and made major legislative efforts to improve the fairness and quality of American life. But he served in turbulent times, and neither his presidential campaign nor his presidency were free of some of the baser aspects of political life.

1964				
Lyndon B. Johnson Hubert H. Humphrey	Texas Minnesota	Democratic	486	61%
Barry M. Goldwater William F. Miller	Arizona New York	Republican	52	29%

Lyndon Baines Johnson

Lyndon Johnson lived from 1908 to 1973 and served as our 36th president from 1963 to 1969. Lyndon was born in a community called Stonewall in the Texas hill country. His father, Sam Johnson, farmed and attempted entrepreneurial pursuits in cotton, stocks and real estate. He was occasionally successful. Of his father, Lyndon observed, "When he had too much to drink, he'd lose control. Sometimes he'd be lucky and make money, but more often he lost." Sam Johnson did have a reputation for honesty and was elected to the state legislature. When Lyndon was thirteen years old, his father's investments failed and the family fell into poverty. This was difficult for Lyndon, and he determined to do better. The experience gave him empathy for those in poverty, as well as a desire to help others.

After high school graduation, Lyndon did not know what course to set, so he traveled and did odd jobs. He returned home, stating he wanted to spend his life working with his mind and not with his hands. He enrolled in Southwest Texas State Teachers College and earned a degree while working in an elementary school whose student body was mostly Mexican. At the elementary school, he stressed academics but also used his own money to purchase bats and balls for the students. He is said to have been remembered as "the Anglo who cared."

After receiving his bachelor's degree in 1930, Johnson taught at a high school but realized he had greater ambitions. He later confessed to historian Doris Goodman that he had never wanted to be an engineer, banker or businessman. "The thing that gives me the greatest satisfaction is dealing with human beings and watching the development of those human beings." He entered politics to fulfill that ambition. He served in Washington, D.C. as secretary to Congressman Richard Kleberg from 1931 to 1935. In 1934, he married Claudia Taylor, better known as Lady Bird, and in 1935 returned to Texas to serve as director of Franklin Roosevelt's National Youth Administration. He had a reputation for being an effective manager and for compassion toward black and Mexican-American youths.

In 1937 Johnson, typically known as "LBJ," campaigned as a "New Dealer" and won his own congressional seat at the youthful age of twenty-nine. He lost a bid for the U.S. Senate in 1941 and he is reported to have allowed voter fraud, as did his opponent, in his attempt to win. During the campaign, he promised voters that he would join the fighting forces if America were to go to war.

When the U.S. entered the war, he obtained an appointment inspecting military bases, a post that kept him out of combat. Political pressure motivated him to get closer to the action, so he went to the Pacific as an observer. He accompanied a crew on a bombing run. His plane managed to get back to base after being hit by enemy fire. In a politically motivated gesture, General MacArthur awarded Johnson a Silver Star, though no one else on the flight received that honor. Johnson may have exaggerated his combat involvement and always wore a small bar on his lapel.

Johnson's wife Lady Bird purchased radio station KTBC in Austin, Texas. Lady Bird obtained what observers called unusually quick and favorable Federal Communications Commission approvals. The station was very successful and Johnson's wealth soon far surpassed that of his congressional salary. Johnson insisted, however, that he had no interactions with the business.

Johnson was elected to the U.S. Senate in 1948 in a closely-contested election against former governor Coke Stevenson. He won by an 87-vote margin, but only after a late rally from precinct 13 along the Rio Grande erased his 157-vote deficit. In his book *The Years of Lyndon Johnson: Means of Ascent,* historian Robert Caro asserts that Johnson rigged the precinct 13 results.

Regardless of the means by which he may have been elected, Johnson is regarded as having been one of the most effective senators in history due to his unusual ability to induce people with competing ideas to work together. Johnson became Kennedy's vice president in the election of 1960.

Johnson ascended to the presidency when John F. Kennedy was assassinated in Dallas, Texas on November 22, 1963. At his inauguration, Johnson vowed to continue Kennedy's initiatives. Johnson's presidency began with promise. He made major efforts to pass legislation to help promote what he termed "The Great Society," which he described as "a place where men are more concerned with the quality of their goals than the quantity of their goods." Johnson was firmly committed to equality and Congress passed civil rights laws under his leadership. He passed Medicare reform to help senior citizens with medical bills. He waged a "war on poverty" aimed at improving conditions in inner cities. Johnson chose JFK's brother-in-law, Sargent Shriver, to handle details of the war on poverty, while Johnson used his political skills to get the measures passed. In total, Congress authorized ten separate anti-poverty programs. These included the Job Corps, Volunteers in Service to America (VISTA), the Community Action Programs (CAP), and Head Start.

Johnson was a person of complexity and contradictions. In terms of personal conduct, Johnson was given to backcountry crudeness. He is reported to have met with advisors while seated on a toilet. While giving *Time Magazine* reporters a tour of his ranch in 1964, he reportedly drove ninety miles per hour to purchase beer. However, he had a genuine desire to use his political power to advance the public good.

Johnson's successes and popularity did not last. He may have been a victim of the firmly entrenched American foreign policy of containing the spread of communism. Dwight Eisenhower and JFK had sent military advisors to Saigon to help South Vietnam's defense against communists from the north. The war was going badly and Johnson sent American soldiers to fight alongside the Vietnamese. The U.S. became mired in a war that increasingly appeared impossible to win, and this war would consume Johnson and the nation. LBJ was advised by former President Dwight Eisenhower to win the war. Johnson deepened U.S. military involvement and hid his strategy from the public. This was the most unpopular war in American history and there was growing mistrust of government. There were conflicts between supporters of the war, termed "hawks" and those calling for peace, termed "doves." Some young people burned their draft cards or moved to Canada due to their mistrust of government and their conviction that the war was unjust. The United States appeared more divided and conflicted than it had been since the Civil War. Well ahead of the election of 1968, Johnson stated he would neither seek nor accept another nomination for President.

Johnson's popularity was seriously eroded by the time his term ended. His health had always been uncertain and he was not well when he retired from office. He spent his remaining years at his beloved ranch in Texas, tending to his investments, preparing his memoirs, and overseeing development of his presidential library.

The University of Texas at Austin hosts the presidential library as well as the Lyndon Johnson School of Public Affairs, one of the premier public policy schools in the nation. Johnson died on January 22, 1973, one day before the conclusion of the Paris Peace Accords that ended the Vietnam War. As was the case with Herbert Hoover before him and Jimmy Carter after him, Johnson is an example of a skilled and well-intentioned President whose presidency and reputation were damaged by circumstances.

1968				
Richard M. Nixon Spiro T. Agnew	California Maryland	Republican	301	43.4%
Hubert H. Humphrey Edmund Muskie	Minnesota Maine	Democratic	191	42.7%
George C. Wallace Curtis LeMay	Alabama California	American Independent	46	13.5%

1972				
Richard M. Nixon Spiro T. Agnew	California Maryland	Republican	520	61%
George S. McGovern Sargent Shriver	South Dakota Maryland	Democratic	17	8%
Joseph Hospers Tonie Nathan	California Oregon	Libertarian	1	

Richard M. Nixon

Richard Nixon lived from 1913 to 1994 and served as our 37th president from 1969 to 1974. He resigned from office in the summer of 1974 to avoid impeachment proceedings after the Watergate scandal that occurred during his reelection campaign of 1972. Nixon is the only President to have resigned from office.

Richard Milhouse Nixon was born in Yorba Linda, California. His mother, Hannah Milhous Nixon, was a practicing Quaker and his father, Frank Nixon, converted to the Quaker faith upon marriage. Frank was said to be loud, volatile and argumentative in contrast to his mother, who was reticent, calm and compassionate. When Richard was nine years old, his father moved the family to Whittier, where he built a gas station that provided a middle-class lifestyle. The future president worked in his father's store, where the principal lesson he gleaned was "winning means everything." Frank ruled his family through fear, and young Richard is said to have thus learned that power derives from fear. (This would become evident during his political career, as evidenced by his participation in the U.S. Senate on Joseph McCarthy's Government Operations Committee).

Young Nixon was a serious student unusually adept at memorization. He won a speech contest in high school, and in his speech, he opined that laws must be obeyed "for they have been passed for our own welfare." He enrolled at Whittier College where he assumed many leadership roles. He served as president of his freshman class, his fraternity and the history club. He received a B.A. in history from Whittier College in 1934 and won a scholarship to Duke Law School. He was not well-off financially and lived a very modest lifestyle while studying law at Duke. Although he was elected president of the Duke Student Bar Association, he was known to be reclusive and humorless. Although he graduated third in his class, he was unsuccessful in securing employment at a New York law firm, so he returned to Whittier to practice law. He married Catherine "Pat" Ryan in 1940 and the couple raised two daughters.

He was on active duty in the Navy Reserve in World War II and left the military in 1946 with the rank of lieutenant commander. A banker proposed that he run for Congress. Nixon exhibited a dislike both for the New Deal and labor unions. He decided to use the growing "red scare" to his advantage and accused his opponent of being a communist sympathizer. In *Nixon: The Education of a Politician*, biographer Stephen Ambrose described Nixon as a "brash upstart" who waged a dirty campaign with "a vicious, snarling approach." Nixon was elected to the House of Representatives in 1946, where he served on the House Un-American Activities Committee. He was elected to the Senate in 1950, again using "red scare" tactics against his opponent. In the Senate, Nixon joined Senator McCarthy's Government Operations Committee. During this troublesome and divisive era, hundreds of Americans were suspected or accused of being communists or communist sympathizers. Many lost their jobs or were "blacklisted," effectively barring them from employment. McCarthy's committee subjected these individuals to aggressive and damaging examinations. Many Americans forever harbored a huge mistrust of Nixon due to his having been involved in these nefarious activities.

Dwight D. Eisenhower chose Nixon as his running mate in 1952. Nixon's presence on the presidential ticket hurt Eisenhower's campaign when it was discovered that Nixon had accepted cash for a slush fund to pay for his personal expenses. As Nixon had vociferously accused Democrats of corruption, the scandal made him vulnerable to epithets such as "hypocrite" and "crook." Despite this scandal, Eisenhower was elected and Nixon served as his vice-president from 1952 through 1960. In 1959, Nixon traveled to the Soviet Union where he and Premier Nikita Khrushchev engaged in a debate on the relative merits of communist versus capitalist societies. Nixon performed well in the debate, emphasizing Americans' "right to choose."

Nixon was defeated in a bid for president by John F. Kennedy in 1960. In 1962, he lost a bid to become governor of California, and most observers felt that this signaled the end of his political career. In his parting comments, he remarked that this was his last press conference, famously adding "You won't have Dick Nixon to kick around anymore." In 1968, however, he made a remarkable comeback and won the Republican nomination for President. Although he was distrusted by many Americans, he was able to make a good case for himself, promising to be a force for unity. Despite several startlingly insensitive comments made by his running mate, Maryland Governor Spiro Agnew, Nixon won the presidency, defeating Democratic nominee Hubert Humphrey.

Nixon ascended to the presidency amidst the turmoil created by the protracted, unpopular and unsuccessful war in Vietnam. At first, Nixon intensified the war, ordering the secret bombing of Cambodia, a neutral nation harboring communist bases. Pilot records were falsified to cover up this unpopular tactic. Nixon considered the use of nuclear weapons to devastate the North Vietnamese countryside, but demonstrations in the United States were rapidly growing, and he realized he lacked support for such a destructive path. It became increasingly clear that victory was unobtainable so, in 1973, with the slogan "Vietnamization," he withdrew our troops.

Prior to 1969, individuals or corporations who received most of their income from sources such as capital gains, accelerated depreciation on real estate, or other tax-preferred activities paid taxes at a low and presumably unfair rate.

The Tax Reform Act of 1969 was a federal tax law signed by President Richard Nixon. Its largest impact was creating the Alternative Minimum Tax, intended to tax high-income earners who had previously avoided taxes due to various exemptions and deductions. The Act also regulated abuse of dollars put into charitable activity by the rich. For instance, the Act prohibited officers from "self-dealing" – from financially benefitting from their business transactions with the foundation. Further, foundations were required to refrain from attempting to influence legislation, and to pay a 4 percent tax on investment income.

Tax laws under the 40th, 43rd and 45th U.S. presidents would greatly expand the national debt by giving huge tax breaks to those who were already very wealthy. In contrast, tax reform under Nixon was intended to make the tax laws fairer.

Nixon had some foreign policy successes worthy of recognition. Most noteworthy was his successful effort to open diplomatic relations with China. Dr. Henry Kissinger had served as Nixon's foreign affairs advisor when Nixon was in the Senate, and was his Secretary of State when he was President. When serving in the Senate, Nixon may have set the stage for establishing relations with China. In the magazine *Foreign Policy*, he wrote, "There is no place on this small planet for a billion of its potentially most able people to live in angry isolation."

During his first year in office Nixon sent a private message to the Chinese, expressing his desire for a closer relationship. The process proceeded in small steps. It was considered a breakthrough when Chairman Mao Zedong invited American table tennis players to go to China to play against Chinese players. Nixon followed by sending Kissinger on a secret mission to meet with Chinese officials. The world was stunned when, on July 15, 1971, the United States and China simultaneously announced that President Nixon would visit China. In February 1972 President Nixon, accompanied by his wife Pat, Secretary of State Kissinger and 100 journalists, went to China where Nixon and Kissinger met with Chairman Mao and with Premier Zhou Enlai. The visit ushered in a new era of communication between the two countries and was undoubtedly the greatest achievement of Nixon's presidency.

In July of 1970 President Nixon proposed the establishment of the Environmental Protection Agency. The EPA began operation in December of 1970, after Nixon signed an executive order. The

House and Senate later ratified the establishment of that agency. Nixon's message to Congress included the following:

> Each of us across this great land has a stake in maintaining and improving environmental quality. Clean air and clean water, the wise use of our land, the protection of wildlife and natural beauty, parks for all to enjoy. These are part of the birthright of every American. To guarantee that birthright we must act and act decisively.

The Environmental Protection Agency remains to this day a major force for conservation. Other events may have caused people to forget this legacy of Nixon's presidency.

In some ways, Nixon appeared to behave more like a king than like the president of a democratic nation. He would host soirées at the White House and, with all the company assembled, he and his wife Pat would descend a staircase as a band played *Hail to the Chief*.

Nixon did not always surround himself with persons dedicated to honesty, fair play and high ideals. Many of his associates engaged in what became known as "dirty tricks." These methods included harassing activist groups and putting surveillance devices into the offices of opponents.

Politicians often employ aggressive, dishonest, or underhanded means to get into office, intending to do good for society once in office. It is a Machiavellian principle: "the end justifies the means." Since many politicians use such means to attain their office, perhaps they can best be judged by whether they forsake illicit practices and employ more honorable methods once in office. In the case of Richard Nixon, he employed "dirty tricks" to attain office, but would not forsake those methods.

When five men were caught breaking into Democratic headquarters in the Watergate complex in Washington, DC, a large-scale investigation began. Carl Bernstein and Robert

Woodward, investigative reporters employed by the Washington Post, were instrumental in revealing the extent of the Illegal and underhanded tactics that had been used by the Committee to Re-elect the President. The administration was not sufficiently cooperative with Special prosecutor Archibald Cox. Nixon fired Cox and appointed Leon Jaworski. In the meantime, Nixon's vice president, Spiro Agnew, resigned and was convicted of bribery, tax evasion and money laundering stemming from his tenure as governor of Maryland. Nixon chose Gerald Ford as Agnew's replacement.

Evidence of a cover-up mounted and support for the President waned. Nixon faced allegations of improper use of government agencies, accepting gifts while in office, and tax evasion. Nixon had arranged for all oval office conversations to be taped, which led to all of America hearing him direct his subordinates to use agencies such as the IRS "against our enemies, not our friends." The level of corruption in Nixon's administration was significant. On May 9, 1974 the House Judiciary Committee opened impeachment hearings and on August 9, 1974 Nixon resigned.

Gerald Ford succeeded Nixon and on September 8, 1974 granted Nixon a "full, free, and absolute pardon." This was a controversial decision, as many observers insisted that Nixon deserved indictment for his actions. During his twenty years of retirement, Nixon did his best to rehabilitate his image. He wrote his autobiography and nine other books.

In 1980 Nixon published *The Real War*, advancing tough-minded views of how America should stay militarily strong to avoid an eventual war of annihilation. The book was received with very divided opinion. There was a more positive response to *Leaders: Profiles and Reminiscences of Men Who Have Shaped the Modern World*, published in 1982. In that book, Nixon made observations about the character and policies of leaders he had met while in office.

234

Nixon suffered a debilitating stroke on April 18, 1994 and died four days later at age 81. The Watergate debacle had been an enormous political scandal. The breadth of violations of law and ethical conduct, as well as the breach of public trust, cannot be overstated. Corruption and Watergate tainted history's view of Nixon's presidency. Nevertheless, the establishment of the Environmental Protection Agency and the opening of relations with China remain achievements of major significance.

Events or Themes in Era 208

<u>The Civil Rights Movement</u>

This is a topic that by itself could fill many volumes. The topic cannot be fully developed here, but to leave it out would be inappropriate. The following is a summary of a few of the events of this era.

The Civil War and the 14th and 15th amendments may have ended slavery as an institution, but they did not end the suffering of black Americans, particularly in the southern states. In 1868, equal protection under the law was established by the 14th Amendment to the Constitution. In 1870, the 15th Amendment granted blacks the right to vote. Nevertheless, many white people, especially in the South, were unhappy. People whom they had once enslaved were now their legal equals. Racial segregation was widely practiced in southern states, with black Americans being obliged to attend different schools, use separate public restrooms and to sit in the back of public buses. Those are but a few of the injustices and indignities to which black Americans were subjected.

Early in the Reconstruction era, blacks took on leadership roles they had never before had. They held public office and sought legislative changes for equality and the right to vote. However, during Reconstruction, the desires of Presidents Lincoln and Johnson for reconciliation without punishment were not realized. The South was subjected to military occupation and political injustice, and the southern states retaliated with so-called "Jim Crow Laws." Those laws restricted the rights of black Americans as noted above. In addition, interracial marriage was illegal, and many blacks could not vote because they were unable to pass voter literacy tests. Moreover, southern segregation gained ground in 1896 when the U.S. Supreme Court declared in Plessy v. Ferguson that facilities for blacks and whites could be "separate but equal."

In the 1950s and 1960s the Civil Rights Movement was a prominent part of the American social and political landscape, and there were landmark events. One such event was the enrolling of two black students at the University of Alabama. Indicative of how much support there was for racial segregation was Alabama Governor George Wallace's attempt to physically prevent the two black students from enrolling at the University.

Following his election as Governor of Alabama, George Wallace had famously stated in his January 1963 inaugural address: "segregation now, segregation tomorrow, segregation forever." On June 11, 1963, when black students Vivian Malone Jones and James A. Hood showed up at the University of Alabama campus in Tuscaloosa to attend class, the governor literally stood in the doorway as federal authorities tried to allow the students to enter. President John F. Kennedy called for troops from the Alabama National Guard to assist federal officials, and Wallace stepped down. Vivian Malone Jones became the university's first black graduate.

Since that era, there has been a multitude of advances in civil rights. Many more persons of color have made their way to prominence in such areas as law, politics, sports (think quarterbacks, coaches and managers), science, education, journalism and broadcast news. Although the job is not complete, and there is still prejudice to be corrected, much of the advance that has been achieved has its roots in the civil rights movement of the 1950s and 60s.

Labor Unions Reach their Height of Power

The history of labor in America is, as are many topics briefly summarized in this book, an enormously complex history with ups, downs, twists and turns. The main theme I want to emphasize is that of "pendulum swings." There seem to be many trends in the natural world as well as in human affairs in which moderate, steady states are not common. Rather, trends seem to go from one extreme to another. Ice ages alternate with warmer interglacial periods. Populations of animal species alternate between over-population and extinction or near extinction. Nations alternate between dominance of their era and loss of power. So it has been with labor in America.

Prior to the National Labor Relations Act of 1935, labor was at the mercy of employers. An example is an event known as "The Ludlow Massacre," which occurred in 1914 at Colorado Fuel & Iron, a mining company owned by the Rockefellers. Many workers were living in a makeshift tent city, having been evicted from company housing by the company management. Rockefeller and the company became infamous because twenty people died in a battle between the Colorado National Guard and striking miners. The deceased included ten women and two children, who burned to death after machine gun fire ignited the tent city. At least sixty-six people reportedly died in the open warfare between labor and mine operators in Colorado between spring and fall of 1914; the violence only ended when President Wilson sent Federal troops to the area.

Another example is the 1892 strike at Andrew Carnegie's Homestead, Pennsylvania steel plant. The plant employed 3,000 workers, most of whom were unskilled and paid fourteen cents/hour for a ten-hour workday, 6-7 days per week. In 1889 Carnegie appointed Henry Frick as plant manager. Carnegie's plan was for Frick to sign no new union contract, to overproduce prior to the contract expiration, and to then close the plant to break the union. The union struck in protest and Frick hired strikebreakers and Pinkerton detectives. Gunfights ensued in

July, deaths resulting. The strike and violence continued. By October, sixteen hundred workers were on relief rolls. The union surrendered in November and, of 3,000 original workers, only 800 were re-hired. That ended unions in the steel industry for forty years. Union workers were clearly powerless.

In the 1960s the national labor relations board regulated collective bargaining more than in the past. Resistant corporations were forced to take the process seriously. Unions became very powerful in several industries, famously in steel, auto manufacturing, freight and cargo handling, and construction. By the mid-1950s, unions in the U.S. had successfully organized approximately one out of every three non-farm workers and wages increased dramatically. A pendulum swing had occurred. Wages, once suppressed unfairly, were now higher than were warranted either by the cost of living or by the education or skills of many workers. The power of the unions created those wage structures. Union contracts typically called for grievance procedures and union officials, including union stewards, were allowed company time, and in some cases full-time positions, to tend to union business. Many observers saw union members as reveling in their newfound power to the point of arrogance. Knowing that their unions would protect them from disciplinary action or firing, some union workers gave less than optimal effort to production or to company efficiency. The 1960s represented the peak of labor's power and a peak in the earning capacity of many unionized workers.

For a variety of reasons, the power of unions, as well as union membership, began to drop. Some heavily unionized industries began to lose profitability. Government support of unions fell off, as evidenced by President Reagan's actions in 1981. On August 5 of that year, Reagan fired 11,345 striking air traffic controllers and banned them from federal service for life. Later that year their union was decertified by the Federal Labor Relations Authority. It is estimated that in 1983 the percentage of American workers in unions had dropped from its high of about 33 percent to 20 percent. And, in 2013 union membership was down to 11.3 percent. Significantly, the 2013 figures include 6.7 percent for the private sector and 35.3 percent for the public sector. The swing in power between companies and their workers is noteworthy. For many decades, employers exploited their power over labor. When workers gained strength through their unions, they appear to have exploited their newfound power. Fairness, reason and cooperation appear to have been largely absent during these power swings.

Space Exploration

<u>The 1969 Moon Landing</u>

On May 25, 1961 President Kennedy made the following statement to congress: "I believe this nation should commit itself to achieving the goal, before this decade is out, of landing a man on the moon and returning him safely to Earth." What became the Apollo program was born. Kennedy did not live to see this goal attained.

The Apollo missions preceded in a series of steps. In 1966 the National Aeronautics and Space Administration (NASA) tested the spacecrafts in an unmanned mission. Command and lunar landing modules were built, and were tested in Earth orbit (Apollo 7 and 9) and in lunar orbit (Apollo 8 and 10). Apollo 10 was the dry run for the July 1969 Moon landing, as a crew of three astronauts rode the first complete Apollo spacecraft around the moon.

Finally, on July 20, with Michael Collins piloting the command vehicle in lunar orbit, Neil Armstrong And Buzz Aldrin descended in the landing vehicle and became the first human beings to set foot on another world. They remained on the lunar surface for just over twenty-one hours. Apollo 11 was a great symbolic event and a technological triumph.

Scientific progress would be made on six subsequent Moon missions (Apollos 11, 12, 14, 15, 16, and 17). Data was collected on magnetic fields, soils, meteoroids, seismic activity, heat flow, and the solar wind. 180 pounds of moon rocks were returned to Earth for analysis. The first Moon landing remains a major landmark in space exploration and, perhaps, in humankind's ability to envision our future.

Inventions that Changed Our Lives in Era 208

The LASER

LASER is an acronym that stands for "light amplification by stimulated emission of radiation." The idea of the laser has its roots in Albert Einstein's concept of "stimulated emission of radiation," which he put forth in 1916. It was forty years before the concept was pursued for technological use. Two physicists, Charles Townes and Arthur Schawlow, published a paper on laser theory in 1958. After the publication of their paper, physicists across the world raced to make Townes and Schawlow's theory a reality. The government poured money into public and private laboratories to develop the technology.

It was solo researcher Theodore Maiman at Hughes Electric Corporation in California who created the first working laser, a breakthrough in the field of applied physics. By flashing white light into a cylinder of ruby, Maiman energized electrons in the chromium. The light pulse of the ruby was amplified to high power, resulting in a laser. Physicists were stunned when Maiman published his results in the British journal *Nature* in 1960.

Maiman's successful test ignited a laser boom. The first commercial laser hit the market in 1961. Fast on the heels of Maiman's laser came an array of laser technologies. Lasers are used in a wide and expanding variety of applications, including cutting, drilling, welding, marking, engraving, micromachining, lithography, metrology, alignment, eye surgery, dermatology, and dentistry. Focused lasers have been developed to create a less invasive means of eradicating cancer cells. The laser was one of Era 208's great advances.

The Origin of the Internet

Research into the origin of the Internet reveals a confusing array of agencies, persons, computer systems, and protocols. The internet may have entered its infancy in 1969. Scientists working for the U.S. Advanced Research Projects Agency (ARPA) at the University of California and the Stanford Research Institute connected their computer networks (the ARPANET). At the same time, networks like ARPANET were being created in other countries

To join independent networks into a "network of networks," a set of standard communication devices – or protocols – had to be created and agreed upon. Cooperation between the ARPANET project and international working groups led to the development of protocols, which were called the Internet Protocol (IP) and Transmission Control Protocol (TCP).

In the early 1980s the National Science Foundation (NSF) funded supercomputing centers at universities in the United States. In 1986 the NSF provided access to the supercomputer sites. Called NSFNET, it was available for research and academic organizations in the U.S. and internationally. Commercial Internet service providers (ISPs) began to emerge in the late 1980s.

The Internet allows access to vast amounts of information from a device such as a personal computer, whose storage capacity is tiny in comparison. We tend to use the terms "Internet" and "World Wide Web" interchangeably, but they are not the same. The Internet is a huge network of computers. The world wide web is a storehouse of multimedia in the form of webpages. You need access to the internet to use the world wide web.

We may wonder where the Internet is located. While some data is stored in the hard drives of users, much Internet data is stored in huge data warehouses that occupy thousands of square feet and house thousands of servers. Some data centers belong to well-known internet giants like Amazon, Facebook, and Google, but larger facilities are owned by companies with unfamiliar names. The energy and cooling demands for a data center are tremendous.

The Internet originally served government, industry and academic institutions. Since 1994 it has expanded to a multitude of users, and it has forever changed the way we communicate and transact business. By 2004 there were about 2.4 billion users, which was thirty-seven percent of the world population, and in 2020 there were 4.5 billion users, or sixty percent of the Earth's human population.

Medical Advances in Era 208

<u>The Polio Vaccine</u>

Poliomyelitis is a disease that attacks the nervous system and has affected humanity throughout recorded history. Initial symptoms vary from mild, flu-like symptoms to life-threatening paralysis. Polio can result in paralysis of the arms, legs or diaphragm. Between two and five per cent of people who develop paralytic polio will die. Half of those who survive will have permanent paralysis. The aftermath is often worse than the initial infection. Growth in legs can be slowed. Years after the initial infection there can be new symptoms of weakness, joint and muscle pain, and fatigue.

Since the virus is easily transmitted, epidemics have been commonplace. The first major polio epidemic in the United States occurred in Vermont in the summer of 1894, and by the 20th century thousands were affected every year. In 1952 – an epidemic year for polio – there were 58,000 new cases reported in the United States, and more than 3,000 died from the disease.

242

Although children, and especially infants, were among the worst affected, adults could also be afflicted. Future president Franklin D. Roosevelt was stricken with polio in 1921 at the age of 39 and was left partially paralyzed.

Early in the 20th century, treatments were limited to quarantines and the infamous "iron lung," a metal coffin-like contraption that aided respiration. On March 26, 1953, American medical researcher Dr. Jonas Salk announced he had successfully tested a vaccine against the virus that causes polio. In 1954, clinical trials began on nearly two million American schoolchildren. In April 1955, it was announced that the vaccine was effective and safe, and nationwide inoculation began. Dr. Salk was celebrated as one of the great doctor-benefactors of his time.

New polio cases dropped to under 6,000 in 1957, the first year after the vaccine was widely available. In 1962, an oral vaccine developed by Polish-American researcher Albert Sabin became available, making distribution of the polio vaccine much easier. Currently, there are just a handful of polio cases in the United States every year, most of which are "imported" by Americans from developing nations. A disease that once devastated the lives of millions has been all but eradicated.

First Organ Transplants

The first successful kidney transplant may have been performed on Dec. 23, 1954, by Dr. Joseph E. Murray's team at the Peter Bent Brigham Hospital in Boston. Murray's team transplanted a kidney from a 23-year-old man to his identical twin, whose kidneys were failing. The recipient survived for nine years

The world was astonished when in 1967 South African surgeon Christiaan Barnard replaced the diseased heart of a dentist with that of a young accident victim. The human body will, of course seek to reject foreign tissue, so immunosuppressant drugs were needed to prevent rejection of the transplanted heart. This does, of course, make the recipient more prone to infection. In this case, the immunosuppressant drugs did prevent rejection, but the patient died of pneumonia less than three weeks later. Nonetheless, this was an astonishing achievement and organ transplants would become common within a few decades.

What we Learned about the Universe in Era 208

<u>The Ocean Floor</u>

Earth scientists had traditionally favored the concept of uniformitarianism - the idea that Earth has stayed the same or changed with extreme slowness. They believed the configuration of the continents and the oceans were as they had always been. Although Alfred Wegener had argued in 1912 that the continents move, he was ridiculed. Discoveries during and after World War II vindicated Wegener and changed our understanding of the Earth.

If the continents and oceans had always been the same, then sand and silt washed into the oceans would have settled onto the ocean floors for eons. The ocean floors would thus have been flat and featureless. But ocean mapping via sonar conducted during World War II revealed that the deep-sea floor is not a flat and featureless plain, but rather a dynamic landscape with vast mountain ranges, deep basins, and long trenches.

Exploration of the ocean floors accelerated in the 1950s. It was discovered that there is a great mountain range on the ocean floor virtually encircling the Earth. It is known as the global mid-oceanic ridge, and it is a huge mountain chain over 50,000 miles long. It is often described as winding its way around the globe like a seam on a baseball.

The geological history of the Earth is essentially a history of Earth getting rid of heat from its formation 4.6 billion years ago. The Earth's crust is thinner in the mid oceans, and hot magma rises from beneath Earth's crust to form new crust. This new crust is then forced away from the mid-ocean zone by newer material rising from below. Harry Hess, a Princeton University geologist, and Robert S. Dietz, a scientist with the U.S. Coast and Geodetic Survey coined the expression "seafloor spreading" to describe this process.

Where does all this extra crust go? If Earth's crust expands along the oceanic ridges, it must be shrinking elsewhere. According to plate tectonics, the spreading oceanic crust is forced downward, or subducted, into the Earth along the leading edges of continental crust. Continental crust tends to be lighter than the ocean crust and rides over it. The descending oceanic crust slowly melts at about 400 miles below the surface and becomes reabsorbed. In effect, the ocean floors are continually being recycled with the creation of new oceanic crust and the destruction of old crust. Subduction zones are usually marked by deep ocean trenches that can be six miles deep compared to the ocean's overall depth of 2 to 4 miles. Keep in mind that the Earth's crust moves along at an average speed of about one inch per year. These changes thus require tens of millions of years. The discovery of the true nature of the ocean floor, of the Earth's crust and of plate tectonics has completely altered our understanding of our planet. Plate tectonics is the grand unifying theory of Earth science.

First Discovery of a Neutron Star

Throughout its early existence, a star is a gravitationally-bound nuclear reactor. When the supply of hydrogen in its core is depleted, the core collapses until, if there is enough mass, pressures will be sufficient for the star to fuse helium nuclei into carbon nuclei. Our own star – Sol – will not be massive enough to fuse elements heavier than carbon. But very massive stars will continue to fuse heavier elements until they fuse iron, at which point they contract violently and explode. What happens next depends on the amount of mass that is left. If there are between 1.4 and perhaps 5 solar masses, the star contracts to the point that the very pressure of electrons around nuclei of atoms will not sustain the atoms. Electrons and protons annihilate one another, and the star continues to collapse until it forms a neutron star, a rapidly spinning sphere of neutrons of unbelievable density.

The existence of neutron stars was not proven until Jocelyn Bell discovered the first one in 1967. Bell was a graduate student at Cambridge University who worked with her advisor, Dr. Anthony Hewish, to make radio observations of the Universe. The discovery was made over time with several observations. A neutron star has a mass of at least 1.4 times the mass of the sun but may be about ten miles in diameter. A teaspoon of neutron star material would weigh about 10 million tons. The gravitational field is intense, and the escape velocity is about 0.4 times the speed of light.

<u>Origin of the Elements</u>

Astronomer Fred Hoyle first proposed this theory in 1946, in Era 207 (Economic Recovery, World War II, The Truman Doctrine), and refined the idea in 1954, in Era 208. The idea is called "stellar nucleosynthesis," which is big-word talk for "how stars create elements."

The Universe begins 13.8 billion years ago (13,800,000,000 years ago). In the early Universe, there was only hydrogen, helium, and a trace of lithium. Slap yourself on the shoulder. What you feel there is a great deal more than those three elements. We are made of carbon, hydrogen, oxygen, nitrogen, sulfur, phosphorus, sodium, potassium, calcium, magnesium, iodine, iron, manganese, molybdenum, and about a dozen other elements found abundantly in the Earth's crust. Where did all those elements come from? They came from the cores of stars. That is correct. This is not a myth or a legend, but is well-documented and well-understood science. Our star – Sol – is busily converting hydrogen nuclei into helium nuclei in its core. In about four to six billion years, our star will run out of hydrogen in its core and will begin to fuse helium nuclei into carbon nuclei. That is as far as our star will go. However, stars of much greater mass go much further, and the early Universe was populated by many super massive stars. Those stars go through cycles of fusing heavier and heavier elements, until they fuse iron. As was described in the previous section on the discovery of neutron stars, these stars then contract violently and blow up. You have undoubtedly heard the term "supernova." In such an explosion, all the elements of the periodic table are fused, right up to uranium, and they are scattered into interstellar space.

Over a period of hundreds of millions of years, interstellar clouds of gas and dust may become gravitationally unstable and collapse, forming new stars with planets. Now, however, the stars and their planetary systems have heavy elements – the type of elements of which we are made. Therefore, our Sun is perhaps a fourth or fifth generation star, and we are literally star children. A study of the way the Universe has concocted the elements of which we are made, of the process by which the Earth formed, and of the way in which life has evolved can give us a special and delightful feeling about life, and many of the discoveries that led to that awareness occurred in Era 208.

Origin of Organic Molecules

In June of 1953 a breakthrough experiment was conducted by Stanley Miller at the University of Chicago. Miller was a graduate student working under the auspices of Harold Urey. Miller was interested in the origins of life, and he conducted an experiment that moved origins-of-life research into the limelight. He created a facsimile of what he believed may have been the Earth's atmosphere 3.5 to 4.0 billion years ago. In a tabletop assembly he had a flask containing water, ammonia and methane. He heated the solution and some of it was evaporated into a chamber in which, using electrodes, he simulated lightning. He let the solution sit for a few days and found the water discolored. Analysis revealed that organic molecules had been assembled. This was the first of several revelations that have convinced the scientific world that simple organic molecules are common, and are assembled in a variety of environments. We now know that simple organic molecules can be found virtually everywhere, including in deep mines, in the deep ocean, on meteorites, and even in vast interstellar clouds of gas and dust. At the time, the Miller/Urey experiment was a breakthrough, and it paved the way for increased inquiry into the origins of life.

Another event during Era 208 that would later influence our understanding of life's origins was the fall to Earth of a meteorite on September 28, 1969. It fell to Earth in a cow field in the town of Murchison, about a hundred miles north of Melbourne, Australia. It weighed several kilograms and contained, by weight, 3.5 percent organic molecules. In the 1990s Dr. Dave Diemer of the University of California at Santa Cruz would make some astounding discoveries about the contents of the Murchison meteorite.

<u>Confirmation of the "Big Bang" – Discovery of the CMB</u>

CMB stands for cosmic microwave background. As was discussed in Era 206 (Roaring Twenties, Prohibition, The Great Depression, World War II), Edwin Hubble showed us that the Universe is expanding. But if it is growing larger, in the past the Universe must have been smaller. The logical conclusion is that the Universe must have been tiny and then expanded. This theory was termed "The Big Bang," and was accepted by many astronomers as opposed to the rival theory, called "The Steady State theory." But, if there had been a big bang, would there be some echo of that event? Many astronomers searched for it.

On May 20, 1964, American radio astronomers Robert Wilson and Arno Penzias discovered the cosmic microwave background radiation (CMB). The CMB did not actually originate at the moment of creation. The infant Universe was too dense for photons of light to move freely. About 380,000 years after its creation, the Universe had cooled to the point at which electrons could stick with protons, and the Universe "cleared up." At that point photons were free and the ancient light began saturating the universe. Wilson and Penzias discovered it by accident.

248

Penzias and Wilson experimented with a supersensitive antenna built to detect radio waves bounced off balloon satellites. To measure these faint radio waves, they had to eliminate all interference from their receiver. They removed the effects of radar and radio broadcasting, and suppressed interference from heat in the receiver itself by cooling it with liquid helium. But they found a low, steady, mysterious noise that persisted in their receiver. This residual noise, at a wavelength of 7.35 centimeters, was evenly spread over the sky, and was present day and night. They were sure it did not come from the Earth, Sun, or our galaxy. They checked their equipment, and removed droppings from pigeons that had nested in their antenna. The noise persisted, and they concluded it was coming from outside our galaxy, but they did not know how.

At that time, 37 miles away, astrophysicists Robert Dicke, Jim Peebles, and David Wilkinson at Princeton University were preparing to search for microwave radiation at this same wavelength. Dicke and his colleagues predicted that the Big Bang would have released a tremendous blast of radiation. They reasoned that the radiation would have been redshifted and they calculated the wavelength that should be detectable at the microwave part of the electromagnetic spectrum.

A friend told Penzias about a paper by Jim Peebles on the possibility of finding radiation left over from an explosion at the beginning of the Universe's existence. Penzias and Wilson then realized the significance of what they had detected. Penzias invited Robert Dicke to Bell Labs to look at their antenna and listen to the background noise. Dicke, Peebles and Wilkinson interpreted this radiation as a signature of the Big Bang.

To avoid conflict, they published their results jointly. In 1978, Penzias and Wilson were awarded the Nobel Prize for Physics for their discovery. The Big Bang Theory was confirmed, and it stands as the current theory of the creation of the Universe. Our understanding of the Universe was enhanced.

The First Identified Black Hole

As is often the case in science, the idea of a phenomenon may be conceived long before the science exists to explain or verify it. The concept of black holes has such a history, dating back to studies by Britain's Reverend John Mitchell and France's Pierre-Simon LaPlace in the eighteenth century.

A black hole is not a hole, per se. It is a super dense object whose gravitational field is so strong that the escape velocity is faster than the speed of light. It is what remains after a very massive star has collapsed in on itself, leaving behind an intense gravitational field.

Every massive object has an escape velocity. For example, a spacecraft leaving the surface of Earth needs to go about 7 miles per second, or over 25,200 miles per hour, to escape into space. The escape velocity is determined in part by the distance from the center of gravity. If you started from fifty miles above Earth's surface, the escape velocity would be much less. The more massive the object, the greater the velocity needed to escape. Imagine an object so massive that the escape velocity is over 186,200 miles per hour, which is the speed of light.

A black hole is defined by two features. One is the distance from its center at which the escape velocity is the speed of light. Nothing inside that point can escape, since nothing can travel faster than light, and that position is called the "event horizon." The second feature is known as the "singularity," which physicists tell us is the center of the black hole, a place of infinite density at which space-time has essentially folded in on itself.

It was suspected that these objects could exist based on physics such as Einstein's Theory of General Relativity. But observing a phenomenon is better than just conceiving of it. The first observed object that astronomers accepted to be a black hole is Cygnus X-1, an X-ray source in the constellation Cygnus. Cygnus X-1 was discovered in 1971 and is just over 6,000 light years from the Sun. Cygnus X-1 is one of the most studied astronomical objects in its class. The compact object is now estimated to have a mass about 14.8 times the mass of the Sun. It has been shown to be too small to be any known kind of normal star, or other likely object besides a black hole.

A binary system is a pair of stellar objects in orbit about one another. Cygnus X-1 is a one of such a pair. The other member is a blue supergiant star orbiting at about 0.2 astronomical units, twenty percent of the distance from the Earth to the Sun. Material drawn from the atmosphere of the blue giant swirls around Cygnus X-1 the way water swirls around a drain. This material is called an "accretion disc." Matter in the inner disk is heated to millions of degrees, generating the observed X-rays detectable from Earth. The detection of the first confirmed black hole was a milestone in our understanding of the Universe.

<u>Human Origins</u>

We saw in Era 206 (Roaring Twenties, Prohibition, The Great Depression, World War II) the discovery of the Australopithecines, human ancestors who lived four million to two million years ago. In 1960, Mary Leakey and her son Jonathan found a form of hominid at Olduvai Gorge in Tanzania that they believed was different from and more advanced than the australopithecines. They called it Homo habilis (handy human) because it appeared to be the first human to use tools. More remains of Homo Habilis were unearthed between 1960 and 1963 in the Olduvai Gorge by Mary and Louis Leaky.

<u>The Double Helix of DNA</u>

All the information needed to transmit the blueprints of our bodies is contained in twenty to twenty-five thousand coding genes, which are in twenty-three pairs of chromosomes in the nucleus of each cell in our bodies. Rosalind Franklin and Maurice Wilkins at London's King's College studied DNA by beaming X-rays through crystals of the DNA molecule. This technique, called

X-ray diffraction, revealed images on X-ray film typical of a molecule with a helix shape. James Watson and Francis Crick relied on this data in formulating their theory that the DNA macromolecule must be a double helix, resembling a twisted ladder. The "rungs" of the ladder are composed of bases (adenine, cytosine, guanine, thymine). The uprights are built of alternating sugar and phosphate groups. Watson and Crick published their theory in 1953, and it is one of the most famous scientific discoveries of all time.

Space Exploration in Era 208

<u>The Launch of the Pioneer 10 and 11 Spacecraft</u>

A major player in the advancement of space exploration has been the United States' National Aeronautical and Space Administration, typically known as "NASA." Over a period of decades NASA has sponsored considerable research and has conducted hundreds of unmanned and manned space exploration missions. NASA launched Pioneer 10 on March 2, 1972 and launched Pioneer 11 on April 5, 1973. These two exploratory spacecraft have had an enormous impact on our knowledge of our own Solar System.

On July 15, 1972, nineteen weeks after its launch, Pioneer 10 entered the asteroid belt, a doughnut-shaped area 175 million miles wide and 50 million miles thick. The asteroid belt, material left over from the formation of the Solar System, includes objects ranging in size in size from dust particles to rocks as big as Alaska. Material in the belt orbits the Sun at speeds up to 45,000 mph.

Pioneer 10 made its closest encounter to Jupiter on 3 December 1973, passing within 81,000 miles of the cloud tops of the giant planet. This historic event marked humans' first approach to Jupiter and opened the way for exploration of the outer solar system - for Voyager's "tour of the outer planets." Pioneer 10 imaged the planet and its moons, and took measurements of Jupiter's magnetic field, radiation belts, atmosphere and interior.

These measurements and observations vastly increased humankind's knowledge of the Solar System's largest planet.

Following its encounter with Jupiter, Pioneer 10 traveled to the solar system's outer reaches, studying the solar wind (energetic particles streaming from the Sun) and cosmic rays (fast moving charged particles) entering our portion of the Milky Way. Pioneer 10 made valuable scientific observations of the Solar System until its science mission ended on March 31, 1997.

The Planet Venus Revealed

Venus is the third brightest object in the sky, trailing only the Sun and the Moon. Venus has been called our "sister planet," as it has a similar diameter and distance from the Sun as does Earth. Venus' diameter of 7200 miles is about ninety percent Earth's diameter of 7900 miles. Its distance from the Sun of 67 million miles is about seventy percent of the Earth-Sun distance of 93 million miles. Venus was well-known to ancient civilizations. The Maya had a 260-day calendar that appears to have tracked Earth/Venus conjunctions.

Venus is shrouded in dense clouds, and so hardly anything was known about it until the second half of the twentieth century. Since it is covered by clouds, science fiction writers in the mid-twentieth century often wrote stories about Venus, imagining hot, steamy, lush jungles. But the U.S.'s Mariner 2 probe and the Russian Venera 7 and 8 probes dispelled those notions. It was learned that Venus has a surface temperature of almost 900 degrees Fahrenheit and has a dense atmosphere consisting mainly of carbon dioxide and sulfuric acid. Venus' dense, carbon dioxide rich atmosphere appears to have caused a runaway greenhouse effect, resulting in its hellish surface temperature.

The Restless Earth in Era 208

<u>Hurricane Carol</u>

Hurricane Carol was among the worst tropical cyclones on record to affect southern New England. It developed near the Bahamas on August 25, 1954, and strengthened as it moved to the northwest. Hurricane tracking was not the same in 1954 as it is today. There were no satellites with which to see progress of a storm. Observations from land and from ships at sea were the only means of tracking the storm. While paralleling the Atlantic coast, the storm's strong winds and high seas caused minor coastal flooding and slight damage to houses in North Carolina, Virginia, Washington, D.C., Delaware, and New Jersey.

People in Rhode Island reported that huge waves were crashing on the shore, even while weather reports stated that the hurricane was off shore near Cape Hatteras. The hurricane accelerated north-northeast and made landfall on eastern Long Island, New York and eastern Connecticut on August 31 with sustained winds estimated at 110-mph.

Winds on Long Island damaged 1,000 houses, left 275,000 people without electricity, downed many trees, and left the eastern portion of Long Island isolated. Carol brought strong winds and rough seas to coastal Connecticut, Rhode Island, and southeastern Massachusetts. Throughout the region, about 150,000 people were left without electricity and telephone service. 1,545 houses were destroyed and another 9,720 were damaged. Approximately 3,500 cars and 3,000 boats were destroyed. There were 65 deaths and 1,000 injuries in New England. Overall, Carol caused 72 fatalities and was the costliest hurricane in the history of the United States at the time. Following the storm, the name "Carol" was retired, becoming the first name to be removed from the lists in the Atlantic basin.

Miscellaneous Events in Era 208

Alvin and Ocean Floor Exploration

The planets of the solar system are very difficult to reach and study, but our own ocean floor is just as difficult to explore. Pressures are extreme when you get to depths of a mile or more, and there is total darkness. Before the World War II years, when the ocean floors were considered to be flat and featureless, there was little thought about deep sea exploration. Once it was known that the ocean floor has higher mountains and deeper trenches than do the continents, there was immense interest in further exploration. One result was Alvin, named in honor of Allyn Vine, who was instrumental in the creation of the vehicle.

Alvin is a manned deep-ocean submersible research vessel. It is owned by the United States Navy and operated by the Woods Hole Oceanographic Institution (WHOI) in Woods Hole, Massachusetts. Alvin was commissioned on 5 June 1964. Alvin is launched from a support vessel, RV Atlantis, also owned by the U.S. Navy and operated by WHOI. The submersible has made more than 4,400 dives, carrying two scientists and a pilot, to observe the geology of the ocean floor as well as lifeforms that live with super-pressures and total darkness. Alvin has also been used to explore the wreck of Titanic. Research made possible by Alvin has been featured in nearly 2,000 scientific papers.

We now know there are areas of mid-ocean floor spreading, where tectonic plates are separating, allowing hot material from the Earth's mantle to rise to create ocean ridges. There are also hydrothermal vents, where hot gases escape from beneath, creating a warm, nutrient-filled environment where creatures such as tube worms and giant clams exist. Some researchers believe that life on Earth may have first evolved at hydrothermal vents, protected from meteor and asteroid bombardment at the surface. Alvin has been a huge contributor to our knowledge of our planet and its life.

The Murchison Meteorite

On June 28,1969 a large meteorite weighing more than forty pounds fell to earth near Murchison, Victoria, in Australia. It has become one of the most studied meteorites due to its high percentage by weight of organic compounds. The significance of some of the organics found in this meteorite will be discussed in Era 210 ("9/11" Twin Towers attack, Second Persian Gulf War, Climate Change).

Summary Comments on Era 208

(Years: 1953 – 1974)

Era 208 was an era incredibly full of historical events, technological innovation, medical advances, and new knowledge about the Universe, Earth and human origins. World War II was over, but the Cold War between the United States and the Soviet Union continued throughout the era. The policy of containment – designed to halt the spread of communism – continued to dominate U.S. foreign policy until 2016. This policy led to such events as the Vietnam War, the longest and least successful war ever waged by the United States. The Vietnam war was viewed as a Cold War struggle between the U.S. and a Soviet-inspired regime in North Vietnam. However, this was probably a misunderstanding of the situation. It was fierce nationalism on the part of the Vietnamese, and not Soviet influence, that inspired our adversaries. The Vietnam War essentially spoiled the presidencies of Lynden Johnson and Richard Nixon.

Era 208 saw outstanding medical advances. Polio had been a devastating disease with major outbreaks affecting millions of individuals. Polio has been virtually wiped out with the use of vaccines. The first organ transplants were accomplished, beginning a trend that now saves thousands of lives annually. Laser technology was developed, and this now has scores of applications in several fields from medicine to manufacturing. Space exploration took a major step forward. Though more an engineering triumph than a scientific advance, landing men on the Moon was an event that inspired us about what we could achieve and that stimulated speculation about humanity's future.

The area that saw the greatest advances was science. Our understanding of the Universe, the Earth and human origins all took enormous strides forward. We learned that the ocean floors have higher mountains and deeper trenches than do the continents. We discovered zones of ocean floor spreading and hydrothermal vents where upwellings of heat and nutrients create an ecosystem for exotic creatures such as tube worms and giant clams. We found the cosmic microwave background, evidence of the massive event, termed the "Big Bang," that started our Universe. We identified evidence of massive stars that had exploded in supernova explosions. Those stellar remains include neutron stars and black holes. Our very existence became more fascinating and miraculous as we learned how stars evolve, and how former generations of stars created the elements of which we are made. And we discovered remains of Homo Habilis – "handy man" – a presumed human ancestor who lived perhaps two million to one million years ago.

By the time Era 208 concluded, there had been some progress in racial justice. Though prejudice and discrimination still existed, they were pushed back to a zone of social unacceptability in our public discourse. At least until the 2016 election year, no presidential candidate after Barry Goldwater has ever promoted policies that would place any racial group at a disadvantage.

ERA: 209

Years: 1974 – 1993

(Camp David Accord, Iran Hostage Crisis, Persian Gulf War)

Presidents: Gerald Ford, Jimmy Carter, Ronald Reagan and George H.W. Bush

Gerald Ford

Gerald Ford lived from 1913 to 2006 and served as our 38th president from 1974 to 1977. He was born Leslie Lynch King, Jr., in Omaha, Nebraska. His mother left his father when the future President was an infant. She re-married Gerald Rudolph Ford and began calling her son Gerald Rudolph Ford, Jr., a name he legally adopted at age 22. His stepfather taught him the value of honesty and hard work. The future President played football and earned all-state honors at age eighteen. He attended the University of Michigan where he was a star football player. He turned down offers to play professional football and instead took a job at Yale University as an assistant football coach. His coaching duties interfered with his admission to Yale Law School, so he took summer courses at the University of Michigan Law School and was admitted to Yale Law School in 1938.

While at Yale, Ford was a member of a group determined to keep the United States out of World War II and signed a petition supporting the 1939 Neutrality Act. Ford graduated from Yale Law School in 1941 and returned to Grand Rapids to practice law. However, after the bombing of Pearl Harbor, he joined the navy. He served as an officer on the Monterey, an aircraft carrier that saw considerable action in the Pacific Theatre in 1943-1944. He returned to Grand Rapids in 1946 and stated that his experiences had changed his views. He considered himself "a confirmed internationalist." He was encouraged to enter politics and was elected as a Republican to the House of Representatives in 1949, and he served there for twenty-five years.

Barry Goldwater and his right-wing supporters gained control of the Republican party in 1963, with disastrous results for Republicans in the 1964 national election. Subsequently, Ford led a group of moderates to influence the party to assume a less radical position and, in 1965, the Republicans chose Ford as house minority leader. During his career in the House of Representatives Ford reportedly saw himself as a negotiator and conciliator, but is said to have never written a major piece of legislation. He considered himself to be a fiscal conservative, a moderate in domestic affairs and an internationalist in foreign affairs. Ford served as house minority leader from 1965 to 1973 and was known for being a fair leader with an inoffensive personality. When Richard Nixon became President, Ford found the White House to be too controlling, and complained that Nixon's aides did not respect congressional prerogatives. When Nixon's vice president, Spiro Agnew, resigned amid scandal, Nixon chose Ford to be his new vice president. After Nixon resigned, and immediately after taking the oath of office, Ford addressed the nation and, among his remarks he said, "The Constitution works. Our great republic is a government of laws and not of men."

Gerald Ford became President during an extremely troubled time. The war in Vietnam had ended in what historians agree was a sound U.S. defeat. Faith in government was low and the nation was deeply divided. Although Nixon had resigned, the House of Representatives voted 412 to 3 to accept its judiciary committee report regarding impeachment. In addition, every senior attorney on Special Prosecutor Leon Jaworski's team was in favor of indictment with subsequent trial in the Senate. On September 8, 1974, Ford issued a proclamation granting Richard Nixon an unconditional pardon for any crimes he may have committed against the United States as President. He explained his decision to house majority leader Tip O'Neill, saying he did it "for the good of the nation." An investigative reporter believed that Nixon had offered Ford the vice-presidency on the condition that he be pardoned, and that while Ford was deliberating, Nixon called Ford and threatened to reveal that arrangement if a pardon were not forthcoming. The report of that quid pro quo has not been confirmed.

Eight days later Ford took an additional step that he thought would help heal the nation. He introduced a conditional amnesty program for persons who had evaded the draft during the Vietnam era. He did not extend the unconditional pardon that he had extended to a president who had faced certain impeachment and almost certain conviction. Unconditional pardon for Vietnam era draft evaders would come with President Jimmy Carter's administration.

Inflation was a major problem facing the American economy, partially due to a boycott of the U.S. by the Organization of Petroleum Exporting Countries (OPEC). Ford introduced the "WIN" program – whip inflation now. He encouraged citizens to restrict spending and to wear "WIN" buttons. This program was viewed as window dressing and did not address the underlying problems of the economy. Ford proposed reduced spending and an income tax hike to help with the budget deficit. The Democrats argued that the nation was on the verge of recession, and argued in favor of a tax cut. That judgment turned out to be correct. To try to curb spending Ford used his veto power sixty-six times and was overridden only twelve times.

On the foreign affairs front, President Ford continued efforts to reduce tensions with the Soviet Union and China. In 1975 France invited a group of six nations: France, Germany, Italy, Japan, the United Kingdom and the United States to a conference. Ford secured membership for Canada and attended the so-called G7. He supported international cooperation, and he stated that we live in an interdependent world.

Ford's credibility was low, and suffered even more as a result of a covert action he took in 1975 to try to aid a rebel group in Angola that was fighting against a Soviet-backed movement. Ford's action prompted Cuba to send troops to help the Soviet-backed group, and the conflict thus expanded. When the plan was exposed by the *New York Times*, the Senate investigated, and passed a law requiring Ford to stop his intervention. 1975 was not a good year for Gerald Ford on the foreign policy front. That same year, the Khmer Rouge in Cambodia captured an American merchant ship in the Gulf of Siam. Though he could have negotiated for the release of the ship's crew, which the United States had done in many previous instances, Ford instead listened to Secretary of State Henry Kissinger, who recommended a tougher response. Ford sent aircraft carriers and ordered the landing of a hundred marines. The Cambodians were unexpectedly aggressive, and within an hour eight U.S. helicopters were shot down and twelve marines were killed. The Cambodians released the ship's crew soon after the hostilities began, though it may have been pressure from China and not the attack that prompted it. More marines lost their lives than there were crewmen on the ship. Though Ford called the operation a victory, this interpretation rings hollow.

In the election year of 1976 Ford won the Republican nomination over Ronald Reagan by a very narrow margin. That year, Presidential debates were held for the first time since Nixon and Kennedy had debated in 1960. Ford lost a close election to Jimmy Carter. At Jimmy Carter's inauguration, the newly elected President said, "For myself and my nation, I want to thank my predecessor for all he has done to heal our land."

After leaving the presidency, Ford managed to secure an income for his children via a successful investment in oil with a partner. He continued to appear at ceremonial events such as presidential inaugurals and memorial services. Ford's autobiography, *A Time to Heal*, was published in 1979. A review in *Foreign Affairs* described it as "short, unpretentious and honest."

Ford's successor, President Jimmy Carter, had his senior staff provide Ford with monthly briefs on domestic and international issues, and Ford was always invited to lunch at the White House whenever he was in Washington, D.C. Ford and Carter developed a close friendship after Carter left office, and the two former presidents served as honorary co-chairs of the National Commission on Federal Election Reform in 2001 and of the Continuity of Government Commission in 2002. Until Ford's death, Carter and his wife, Rosalynn, visited the Fords' home frequently. Gerald Ford died on December 26, 2006 at the age of 93 at his home in Rancho Mirage, California.

1976				
Jimmy Carter Walter Mondale	Georgia Minnesota	Democratic	297	50%
Gerald Ford Robert Dole	Michigan Kansas	Republican	240	48%

James Earl Carter

James "Jimmy" Carter was born in 1924 and served as our 39th president from 1977 to 1981. Jimmy Carter was raised in Plains, Georgia in a family that had a modicum of financial success through peanut farming. As a child, Jimmy would often awaken at 5:00 a.m. to help on his father's farm. From an early age Jimmy dreamed of attending the U.S. Naval Academy in Annapolis. He began college at Georgia Southwestern College and subsequently attended Georgia Institute of Technology. He attained admission to the Naval Academy in 1943 and graduated in the top seven percent of his class with a Bachelor of Science degree in 1946.

In 1948 he trained as an officer for submarine duty. In 1952 he worked with Hyman Rickover's nuclear submarine division. Rickover sent Carter to Union College in Schenectady, New York to study nuclear physics and reactor technology. Carter was on assignment with the Nuclear Regulatory Commission when there was a nuclear reactor accident at Canada's Chalk River Laboratories. Carter was ordered to Chalk River to lead a shutdown of a reactor that had suffered a partial meltdown. The team's job was to disassemble the crippled reactor. Team members, wearing protective gear, were lowered into the reactor's core for a few minutes at a time, reducing their exposure. While President, Carter remarked that his views on nuclear energy were influenced by the Chalk River experience.

Carter served in the Navy for seven years as chief engineer on the Seawolf, a prototype nuclear submarine. He was being trained for duty on a more advanced nuclear submarine when his father died. He obtained a release from active duty in 1953 so that he could return to his family's business. He was in the Navy Reserve until 1961. Carter inherited little due to a combination of his father's having generously forgiven debts and the division of the estate among his younger siblings. He set a goal of expanding the family's business and he achieved that goal.

Jimmy Carter is a man of compassion and decency. He wanted to improve the political climate, which was dominated by conflict between racial segregation and the civil rights movement. He refused to join the White Citizen's Council, which attempted to recruit him with the aim of protecting white privilege. Carter was an activist in the Democratic Party. He was elected to the state senate in 1962 after forcing a recount to reverse a malignant case of election fraud. That experience made a deep impression on him, prompting his reflection:

> The events of 1962 opened my eyes not only to the ways in which democratic processes can be subverted, but also to the capacity of men and women of good will to engage the system and right such wrongs.

Jimmy Carter remained true to his values, defending what he deemed just during his political career and beyond. He served in the Georgia Senate for two terms and promoted educational reform. He lost his first bid to be Georgia's governor and that defeat resulted in a period of depression. During the 1970 election campaign, he may have been expedient when he intimated to segregationists that he might be sympathetic to some of their views. He was elected governor and resumed a progressive stance. In January of 1971, in his inaugural address, he stated, "No poor, weak or black person should ever again have to bear the burden of being deprived of the opportunity for education, job, or simple justice."

As governor, Carter took the ethical high road. He was unwilling to hand out favors, such as preferential treatment regarding promotions, in order to secure votes for legislation he sponsored. Carter did achieve some impressive victories. He achieved efficiency in state government by combining three hundred state agencies into twenty-two, and he implemented a new budget method. He authorized the opening of one hundred new community mental health centers and increased the number of minority group members in state positions. There was an expression at the time - "The New South" - implying that the South was undergoing transformation. Carter was considered a "New South Governor." As his term as governor drew near its end, Carter announced his intention to run for President.

Carter was a dark horse candidate when the 1976 presidential campaign began. He traveled the country extensively, and his sincere style and inspirational message won supporters. Early in his term as Georgia's governor, the *Atlanta Constitution* had written, "Virtue is certainly on his side, but how many votes has virtue?" In the 1976 campaign, however, the themes of virtue and effective government resonated with a public weary of deceit. Carter won the Democratic nomination and was elected. On his second day in office, to further heal the country's Vietnam era wounds, Carter issued an unconditional amnesty for all persons who had evaded the draft during that time.

Carter took office at an inopportune time. The country was in the throes of inflation and recession, referred to as "stagflation." We were not past the energy crisis of earlier in the decade, and America's dependence on foreign oil had increased from 35 to 50 percent. President Carter took important steps to try to improve the nation. He established two new cabinet-level departments: The Department of Energy and the Department of Education. One of his goals for the Department of Energy was to stimulate development of new technology. He also promoted conservation and price control. Carter signed the Chrysler Corporation Loan Guarantee Act of 1979 that bailed out the sagging corporation by infusing it with 3.5 billion dollars in aid.

Carter had noteworthy activity in foreign policy. The United States had "owned" the Panama Canal via a perpetual lease since 1903, but opposition to U.S. ownership was building. In 1977, President Carter and Panama's National Guard Commander Omar Torrijos signed what was called the Torrijos-Carter Treaties. Under the agreement, the U.S. retained the permanent right to defend the canal from any threat that might interfere with its neutral service to ships of all nations.

Tensions in the Middle East were enormously high, particularly between Israel and Egypt. The fear was that the conflict between those two nations could precipitate a confrontation between the United States and the Soviet Union. Carter invited President Anwar Sadat of Egypt and Premier Menachem Begin of Israel to meet with him at Camp David near Washington. The three worked tirelessly with their advisors to draw up a treaty to achieve a just and lasting peace. Carter played a major role and the result, called the Camp David Accord, was a treaty signed on March 26, 1979. The accord guaranteed peace between Israel and Egypt. The Camp David Accord was a major achievement and may have been the high point of Carter's presidency.

As the stock market crash of 1929 tarnished the presidency of Herbert Hoover, an event was to take place that would have a similar effect on Jimmy Carter's presidency: revolution in Iran.

265

The Shah of Iran, who had been supported by the U.S., fled the country, and a religious leader, Ayatollah Khomeini, took control. Oil shipments to the U.S. were disrupted, and there were shortages and price increases in the U.S. Consumers had to line up at gas stations for an hour or more just to fill up their tanks.

To add to the worsening atmosphere, militants stormed the U.S. embassy in Teheran. The militants held American employees hostage for 444 days. Americans rallied around the President when, in response to the hostage crisis, he froze all Iranian assets held in American banks. But that support soon waned. The hostage crisis persisted and American citizens, accustomed to viewing their country as all-powerful and dominant, were left with a feeling of helplessness for which they blamed Carter.

In April 1980, Carter approved a rescue mission involving helicopters and transport planes. Mechanical problems and a sandstorm caused the mission to fail, compounded by an accident resulting in the deaths of eight soldiers. The U.S. and Carter appeared inept due to the mission's failure.

Jimmy Carter's popularity sank to the lowest level of any sitting President and he lost the 1980 election convincingly. An event that may have been impossible to predict marred the presidency of one of the most intelligent and decent men to occupy the office. Perhaps Carter erred in running for re-election at a time when his popularity was low and defeat appeared inevitable.

After leaving office, Jimmy Carter continued his humanitarian pursuits. In 1982, he founded the Carter Center, a non-profit, non-partisan organization. He became involved with Habitat for Humanity in 1984 and is that organization's highest profile supporter. In 2002, Jimmy Carter received the Nobel Peace Prize for his work "to find peaceful solutions to international conflicts, to advance democracy and human rights, and to promote economic and social development through the Carter Center."

Many persons worldwide, but residents of Africa in particular, had been victimized by an extremely painful and debilitating disease lasting 10 to 14 months and caused by the Guinea worm. In 1986, the Carter Center led efforts to eradicate the disease. In 1991, the World Health Assembly counted 400,000 new cases each year, but in 2018 only 28 cases were reported worldwide, a 99.99 percent reduction.

Most former presidents have made tens of millions of dollars on private-sector opportunities, such as seats on corporate boards and huge fees for speeches. Carter, though, has chosen to live modestly. He explained, "I don't blame other people for doing it. It just never had been my ambition to be rich." Thus, Carter has set a rare example of dedication to the betterment of humanity combined with a lack of desire for personal aggrandizement.

1980				
Ronald Reagan George Bush	California Texas	Republican	498	51%
Jimmie Carter Walter Mondale	Georgia Minnesota	Democratic	49	41%
John B. Anderson Patrick J. Lucey	Illinois Wisconsin	Independent Democratic	0	7%

1984				
Ronald Reagan George Bush	California Texas	Republican	525	59%
Walter F. Mondale Geraldine Ferraro	Minnesota New York	Democratic	13	41%

Ronald Reagan

Ronald Wilson Reagan lived from 1911 to 2004 and served as our 40th president from 1981 to 1989. He was raised in small towns in northern Illinois. His father, Jack Reagan, worked in sales, but his drinking habit impoverished his family. A biographer indicated that young Ronald was probably quite embarrassed about his family's circumstances, and he developed a lifelong habit of rarely revealing much about himself. He played football at Dixon High School and won a scholarship to Eureka College, a school run by the Disciples of Christ. The future President had little interest in academics and was an average student. He played sports, acted in a few plays and, as a senior, was elected student body president.

After graduating from Eureka College in 1932 he worked for a radio station as a sports broadcaster, and later for another station as a sports director. In 1937 he auditioned at Warner Brothers Studios in California. He relocated when he was offered a contact and became a movie actor. He met actress Jane Wyman on a movie set in 1939. They married in 1940 and would have two children. Reagan played the part of football player George Gipp in the movie *Knute Rockne*. He received good reviews and soon was offered better parts.

Reagan obtained a deferment when the U.S. entered World War II. He made films for the military during the war years. Reagan's marriage suffered when his wife did better in her career than he did, and the couple divorced. He was twice president of the screen actors' guild. The country had become gripped with fear and paranoia about supposed communist influence, and Reagan threw himself into the spirit of suspicion and retribution. He was in favor of "red scare" tactics, including the blacklisting of actors, and he went so far as to inform the FBI that he believed that certain Hollywood persons harbored communist sympathies. In 1952 he married Nancy Davis. In 1954 he became host of *General Electric Theatre*, and his job was to promote GE's image. He gave speeches, primarily to right-wing groups, and was harsh and openly critical of American foreign policy.

Reagan considered himself a Democrat until 1962 but became a Republican in 1964. Significantly, he backed Senator Barry Goldwater's presidential bid in 1964, when Goldwater's platform was considered by most observers to be reactionary, racially biased and militarily dangerous. That campaign drew attention to Reagan, who was recruited by wealthy businessmen to run for governor. In his campaign he condemned student protesters and vowed to silence them. His tactics succeeded and he was elected governor of California in 1966.

As governor of California, Reagan's actions were repressive. He criticized university officials for tolerating student demonstrations at the University of California's Berkeley campus. On May 15, 1969 Reagan sent the state highway patrol to subdue The People's Park protest. The incident led to what became known as "Bloody Thursday" due to the shooting death of a student. Subsequently, Reagan called in 2,200 state national guard troops to occupy Berkeley. A year after Bloody Thursday, Reagan was questioned about campus protests and is reported to have replied, "If it takes a bloodbath, let's get it over with. No more appeasement." In addition to his repressive attitude regarding dissent, Reagan was criticized for working short hours and for his distaste for details, but he was re-elected in 1970.

In the elections of 1972 and 1976 Reagan aimed for the Republican nomination for President but fell short. He did become the party's nominee in 1980. In the election of 1980, the Iran hostage crisis had severely affected public morale. Americans had never felt so powerless and frustrated, and many people blamed President Jimmy Carter for the situation. Reagan promised he would cut taxes and balance the budget. George H.W. Bush, his opponent during the primaries, termed that promise "voodoo economics." Nonetheless, Reagan won the nomination and he chose Bush to be his running mate. Reagan won the support of evangelical Christians because he opposed abortion and favored prayer in schools. Their support of him became even greater when Reagan said he doubted the theory of evolution. In that atmosphere Reagan won a decisive victory.

As President, Reagan supported and signed the Economic Recovery Tax Act of 1981. This act gave small tax breaks to lower income citizens and huge tax breaks to wealthy Americans, which Reagan stated would boost the economy, increase revenues and balance the budget. Reagan stated that his plan was well thought out and based on supply-side economics. Reagan's budget director David Stockman, however, knew the plan was flawed. The stock market soared, but in his first term in office Reagan added more to the national debt than had every other President before him combined. This confirmed Stockman's contention as well as the aspersion Bush had cast about Reagan's economic plan. In exchange for the debt, wealthy people purchased unprecedented numbers of large homes, luxury cars and high-end retail goods. These items have done little to defend our shores, renovate our infrastructure, or educate citizens. As this is written in 2022, the country's debt is so high that about one-half of all tax revenues are used just to pay interest on the debt. People recall that the stock market rallied under Reagan and that business did well, but few seem to remember that the spiraling debt problem began with Ronald Reagan's flawed policies.

Reagan weakened enforcement of environmental codes, practically gave oil and coal rights to corporations, and reduced enforcement of health and safety codes. His attitude toward unions was clear when he fired thirteen thousand air traffic controllers who had staged a job action. Overall, his policies were repressive and reactionary.

Americans have enjoyed freedom due in great part to our separation between church and state. Reagan proposed a constitutional amendment that would have weakened this separation, and he repeated his attempt to place prayer in schools throughout his presidency. It should be noted that there has never been a court ruling prohibiting anyone in a school from engaging in prayer in a moment of silence. In fact, as has been noted, "There will be prayer as long as there are math tests." The country is fortunate that Reagan was not successful in weakening our separation between church and state.

The right to vote is another important pillar of American democracy. But voting by itself, though necessary, is not sufficient. The public needs exposure to a variety of ideas, and the Federal Communications Commission's Fairness Doctrine, adopted in 1949, had the purpose of ensuring that the public would be exposed to varying viewpoints. At Reagan's insistence, the FCC discarded the Fairness Doctrine. Although a Democratic Congress passed a bill to reinstate it, Reagan vetoed that bill.

Reagan stated publicly he would never pay to have hostages released. However, in 1985 he agreed secretly to sell arms to Iran for its war with Iraq in exchange for the release of American hostages. When this deal was exposed, it became a major issue known as the "Iran-Contra Scandal." Reagan was obliged to appoint a commission to investigate the affair. It was admitted that documents showing Reagan's intentions had been destroyed. Reagan was thus protected, and responsibility was placed on his staff for arranging the deal and for covering it up. The President was criticized for not having had sufficient knowledge and control of his staff's activities. The Iran-Contra Scandal was a major black mark on his administration.

In the foreign policy arena, Reagan advanced the Reagan Doctrine, which stated that the U.S. would fight communist activity anywhere in the world. One problem is that one person's terrorist is another person's freedom fighter, and it is frequently impossible to tell the difference, or to know what ideology people are fighting for. Nonetheless Reagan managed to get the U.S. involved financially and/or militarily in El Salvador, Nicaragua, Lebanon and Grenada.

On December 26, 1991, the Soviet Union officially granted independence to the republics that had been controlled by the

Union of Soviet Socialist Republics (USSR). Casual observers attribute the break-up of the USSR to the United States' military build-up under Ronald Reagan's presidency. However, the USSR had been falling apart from within. The Soviet command economy had not promoted entrepreneurial activity and had, therefore, stifled innovation and progress. Mikhail Gorbachev, the eighth and final leader of the USSR, wrote the following.

> The wheels seemed to be turning, but nothing much was happening. Something strange was taking place. The huge flywheel of a powerful machine was revolving, while transmission from its drive to the workplace was skidding, or drive belts were too loose.

Gorbachev thus metaphorically described the failure of Soviet industry and the Soviet economy. The Soviet Union was disintegrating. It is unlikely that the military build-up under Reagan was the cause.

In October of 1987 the stock market crashed. The Dow Jones Industrial average fell from 2700 to about 1800 in three days. By the end of Reagan's term, the market was recovering and the Iran-Contra Scandal was receding into the past. Reagan had a soothing voice and a speaking delivery that inspired confidence. At the close of his years in office he remained popular despite a soaring budget deficit, repressive policies, and a scandal in which he very likely committed impeachable offenses.

Ronald Reagan retired from the White House to a home in the wealthy Los Angeles enclave of Bel Air. He spent his time organizing his memoirs and supervising the creation of his Presidential Library in Simi Valley, California. He made headlines for accepting a $2-million fee to speak in Japan, and responded to this criticism by declining to schedule other foreign speeches. Until he contracted Alzheimer's disease, he enjoyed his retirement.

This happy period in his life was soon over. On November 5, 1994, Reagan addressed a letter to the American people. It began: "My Fellow Americans, I have recently been told I have Alzheimer's disease." His faculties had been deteriorating while he was serving as President. In fact, according to his chief of staff Donald Regan, "Every word was scripted, every place Reagan was expected to stand was chalked with toe marks." Without a script, he could not recall specifics and was prone to embarrassing errors. However, Reagan was a trained actor with a soothing voice, and the American people, not adept at due diligence, fact-checking or independent opinion-formation, were deceived.

The letter Reagan had written to the American people spoke of the burden his diagnosis would place upon his wife Nancy, and that he wished for a way to spare her the pain. In the ensuing years, Reagan largely disappeared from public view as Alzheimer's took its toll. The courage with which he and Nancy faced this ordeal raised public awareness and inspired contributions to organizations dedicated to finding the cause and cure for the disease. Ronald Reagan died on June 5, 2004. He was accorded a state funeral and was buried on the grounds of the Reagan Presidential Library. As he had been popular, some historians rate him as having been a good President. However, an examination of his actions both before and during his presidency does not necessarily reflect that view.

1988				
George Bush Dan Quayle	Texas Indiana	Republican	426	54%
Michael S. Dukakis Lloyd Bentsen	Minnesota Texas	Democratic	111	46%

George H. W. Bush

George H. W. Bush, later often referred to as "George Bush, Senior," lived from 1924 to 2018 and served as our 41st president from 1989 to 1993. He was born on June 12, 1924 in Milton, Massachusetts. His father was president of an investment banking firm, served on several boards of directors, and served in the U.S. Senate in the late 1950s and early 1960s. George's mother came from a prominent midwestern family. George attended Phillips academy, a private school in Andover, Massachusetts. He was captain of the baseball and soccer teams as well as senior class president. While a student, he met Barbara Pierce and the couple decided they would marry.

The future President turned eighteen years of age six months after the bombing of Pearl Harbor. He joined the Navy on his eighteenth birthday and became the Navy's youngest pilot. After the war he attended Yale University, where he excelled academically and in athletics. His father arranged a job for him at an oil company in Odessa, Texas. George graduated, re-located to Texas, and was then re-assigned first to California and then to Midland, Texas. At age twenty-seven, he and a partner started their own oil drilling company, and by age forty he had become a millionaire. After founding his own business, he turned his attention to politics. In 1964 he lost in a bid for the U.S. Senate. He was thought of as an outsider by Texans, so he donned cowboy boots and hats and tried to affect more earthy, informal Texas mannerisms. In 1966 he was elected to the House of Representatives and in 1971 President Nixon appointed him Ambassador to the United Nations. In 1973 he chaired the Republican National Committee at President Nixon's request. Bush refused to defend Nixon as the Watergate scandal worsened and, on August 7, 1974 Bush wrote to the President and recommended his resignation.

The following year President Ford appointed Bush as Director of the Central Intelligence Agency. The CIA had come under strong fire for illegal and unauthorized activities. Bush served as

Director for almost a year and is credited with restoring some of the agency's lost credibility. He resigned when Jimmy Carter won the presidency in 1976 and returned to his business interests in Texas. Bush campaigned for the Republican nomination for the presidency in 1980. During his campaign, Bush referred to Reagan's economic plans as "voodoo economics." Reagan won the nomination and chose Bush as his running mate, and Bush ceased his criticism of Reagan's economic policies. Later events confirmed that the criticism had been justified. Bush served as Reagan's vice president from 1981 through 1989. As vice president Bush headed task forces, one of which had the purpose of stopping the flow of illegal drugs into the country. Bush had also served on the National Security Council. Congress considered impeaching Reagan due to the Iran-Contra scandal, and Bush was in the loop. He denied any involvement, and what could have been a crisis for him blew over.

In 1988, when Reagan's two terms were up, Bush's presidential campaign was successful against Democratic nominee Michael Dukakis. Though Bush typically took the high road, his campaign against Dukakis was harsh, negative, and misleading. During the campaign Bush emphatically stated, "Read my lips. No new taxes." Reality sometimes spoils our intentions and our promises, and Bush would later be forced to renege on that campaign promise. The tax increase was probably justified by rising federal deficits. Perhaps it was the campaign promise, and not the tax increase itself, that was unjustified.

At his inauguration, Bush tried to soften the bitter atmosphere of the campaign. He also criticized the material greed of the Reagan years. He extolled the virtue of volunteerism, referring to volunteer persons and organizations as "a thousand points of light." Early in his first term Bush had to confront the rising federal deficit. The deficit had tripled during Reagan's presidency, a sharp reminder of how ill-conceived and misguided "Reaganomics" was. Bush wanted to reduce the deficit through reduced spending. The Democratic majority in Congress favored increased taxes. Bush proposed a compromise that would reduce the deficit through a combination of tax increases and spending cuts, but Republicans resisted. His popularity among Republicans fell and would not recover.

Bush may have made a few missteps in foreign policy. The Soviet Union was dissolving, and Bush may have lost a good opportunity when he did not provide aid to countries trying to establish their independence. In China in 1989 a massive crowd of pro-democracy demonstrators gathered in Peking's Tiananmen Square. Chinese communist leader Deng Xiaoping ordered tanks and heavily armed troops to advance on the protesters, and they fired on or crushed those who tried to block their way. Many demonstrators were arrested and imprisoned. The Chinese government reported 241 killed and 7,000 wounded, though most other sources estimated many more fatalities. Bush publicly announced a suspension of military sales to China, as well as stopping high-level diplomacy. However, he secretly sent an envoy to China to reestablish trade. Americans disliked Bush's soft approach to China's dictatorial rulers.

President Bush signed the North American Free Trade Agreement (NAFTA), which would be ratified and put into effect under President Clinton in 1994. This agreement created a trading bloc of the United States, Mexico and Canada. Though the agreement has come under fire, most economists agree it has been beneficial. The agreement has allowed the United States to benefit from what it does best, which is in the technology and information services sectors. The U.S. may have lost routine, labor-related jobs in manufacturing, but these are areas in which the U.S. had lost its competitive edge.

President Bush's term was marked by military actions in Panama and the Persian Gulf. The nefarious actions of Panamanian leader Manuel Noriega made him a household name in the United States. Noriega had once been supported by the U.S. but was suspected of drug trafficking across the U.S. border and of spying. He also nullified the democratic election of Guillermo Endara to Panama's presidency. President Reagan had tried and failed to remove Noriega from power. President Bush launched an invasion called "Operation Just Cause" and, with 24,000 troops, assumed control of the country. Endara assumed the presidency and Noriega surrendered to the United States. Noriega was convicted and imprisoned in April 1992 for racketeering and drug trafficking.

On August 2, 1990 Sadam Hussein's Iraqi forces invaded and occupied the small nation of Kuwait, thereby threatening Saudi Arabia. The United States encouraged nations on the United Nations Security Council to act to liberate Kuwait. The resolution to use all means necessary to remove Iraqi forces from Kuwait passed the Security Council twelve votes to two with one abstention. Therefore, although the United States led the effort, thirty-five nations were involved. The greatest contributions came from the United States, Saudi Arabia, the United Kingdom and Egypt. Operation Desert Storm began on January 17, 1991 and the coalition forces won a decisive victory. President Bush withdrew U.S. forces without unseating Sadam Hussein. This was controversial, especially as Bush had compared Hussein to Adolf Hitler. Bush may have wanted to withdraw when American casualties were still light. But leaving Hussein in power led to much suffering in Iraq, as well as to a future war.

A Congressional investigation subsequently revealed that Bush's earlier decisions may have led to the war. Bush had provided weapons to Iraq, and his ambassador conveyed American indifference to a potential Iraqi invasion of Kuwait.

Bush sought re-election in 1992, and resorted to negative attacks on his new opponent, Bill Clinton. His attacks carried little weight and opened him up to criticism about the Iran-Contra scandal that had occurred under Reagan. There was mounting evidence from sources such as defense secretary Caspar Weinberger's notes and Israeli intelligence reports that Bush had been involved in the scandal and then lied about it. Bush was soundly defeated in the 1992 election. He retired to Houston, Texas, and four years later no Republican candidate asked for his endorsement.

In retirement, Bush raised money for charities. In 2004 he joined former President Bill Clinton to raise money for victims of the tsunami that had devastated coastal southeast Asia. Clinton and Bush became close friends and raised money for other parts of the world devastated by natural disasters. They teamed up to raise 130 million dollars to help regions in Alabama, Mississippi and Louisiana rebound from Hurricane Katrina. Significantly, in the 2016 presidential election, Bush declared his intention to vote for democratic candidate Hillary Clinton. Bush died in 2018

Events and Trends in Era 209

The Demise of the Fairness Doctrine

To ensure that the voting public would be exposed to a variety of viewpoints on matters of public interest, in 1949 the Federal Communications Commission adopted the Fairness Doctrine. This policy required holders of broadcast licenses to present issues of importance to the public. Importantly, they were required do so in a manner that was—in the FCC's view—honest, equitable, and balanced, with contrasting views presented.

At Ronald Reagan's insistence, the FCC abolished the Fairness Doctrine. The rationale was that the doctrine hurt the public interest by violating free speech rights, but that does not stand up to even casual scrutiny. The true reason the Fairness Doctrine was abandoned was to make the public susceptible to a slanted political agenda. A Democratic Congress passed a bill to re-instate the Fairness Doctrine in 1987, but Reagan vetoed the bill.

The demise of this FCC's Fairness Doctrine paved the way for a television station that, though popular, has contributed to the heightened political polarization in the United States.

The Decline in Power of Labor Unions

In Era 208 (The Cold War, The Vietnam War, The Mook Landing) we saw the rise to a peak of the power of American labor unions.

I also advanced the idea that many trends in the natural world and in human affairs seem to travel from one extreme to another, with moderate, steady states seemingly uncommon. So it appears to have been with the power of labor unions.

The US labor movement was at one time a major force fighting for average workers. Over the last half century, its numbers and power have declined markedly. The share of the private sector workforce that is organized has fallen from 35% to approximately 6.5% today.

The decline gained speed in the 1980s and 1990s, spurred by a combination of economic and political developments. Due to some apparent abuses of union strength, public opinion became somewhat skeptical of the labor movement. Overseas markets increased competition in many industries in which unions held power. Outsourcing became a popular practice among employers. And nonunion firms gained market edge through lower labor costs.

A sharp political turn against labor resulted from President Reagan's public firing of striking air traffic controllers. This vividly demonstrated to a weakened labor movement that times had changed. The pendulum has been swinging back, and organized labor is losing the strength it once wielded.

Inventions that Changed our Lives in Era 209

The First Mobile Phone

American engineer Martin Cooper led the team that built the first mobile cell phone. Cooper worked at Motorola on projects involving wireless communication, such as radio-controlled traffic-lights and the first handheld police radios. Car phones had been introduced by AT&T in 1946, but required much power and were tethered to a car. In addition, there were limited available channels. No portable phones yet existed.

In 1947 AT&T Bell Laboratories showed that dividing a large area into smaller cells allowed more mobile users. When in 1968 the U.S. Federal Communications Commission (FCC) asked AT&T for a plan to employ a portion of the ultrahigh frequency television band (UHF), AT&T proposed a cellular architecture to expand car-phone service. Motorola did not want AT&T to monopolize the industry, and Martin Cooper headed an urgent project to develop a portable cell phone. The result: the Dynamic Adaptive Total Area Coverage phone (DynaTAC), 9 inches tall and 2.5 pounds.

On April 3, 1973, Cooper introduced the DynaTAC phone. To make sure that it worked before the press conference, he placed the first public cell-phone call to the head of AT&T's rival project and gloated that he was calling from a portable cellular phone.

It took eleven more years for Motorola to introduce the first portable cell phone for consumers, but it had a price of $3,995. The cell phones we use today were yet to come, but the basic technology was designed in Era 209.

Medical Advances in Era 209

Magnetic Resonance Imaging

Raymond Damadian is an American physician and inventor of the first MR (Magnetic Resonance) Scanning Machine. Sodium and potassium are crucial to the functioning of every cell in our bodies, and Damadian studied their function. This led to his first experiments with nuclear magnetic resonance (NMR). Damadian discovered that tumors and normal tissue can be distinguished in living tissue by nuclear magnetic resonance, which caused him to propose the MR body scanner in 1969. Damadian invented an apparatus to use NMR safely and accurately to scan the human body, and he was the first to perform a full body scan to diagnose cancer in a human being in 1977.

The method is now well known as magnetic resonance imaging (MRI). It uses a strong magnetic field and radio waves to create detailed images of the organs and tissues within the body. It is a powerful tool for physicians assessing a variety of medical conditions, including tumors, cysts, diseases of the abdominal organs, Injuries of the joints, and abnormalities of the brain and spinal cord.

Increases in Organ Transplantation

The first transplanted organs were kidney and heart. Other organs that have been transplanted include pancreas, liver, lung and intestines. Kidney transplants are the most common type of transplant surgery, with intestines being the least frequent. As transplants became less risky and more prevalent, there was an organ shortage, and the matching and donation process was unwieldy. In 1984 the U.S. Congress passed the National Organ Transplant Act. The act established a national registry for organ matching while outlawing the sale of human organs.

Of course, hearts had to be donated by deceased persons. As organs such as kidneys were donated and transplanted, a person could donate an organ and go on with his or her life. In 2001, for the first time the number of living donors (6,528) exceeded the number of deceased donors (6,081)

In 2005 Baltimore's Johns Hopkins Hospital pioneered the "domino chain" method of matching donors and recipients. Willing donors who are not genetically compatible with their chosen recipients are matched with strangers. In return, their loved ones receive organs from other donors in the pool.

In 2015, over 25,000 people received organs in the U.S. Most recipients were between the ages of 50 and 64. As of January 2016 there were over 121,000 people waiting for lifesaving organ transplants, of whom more than 100,000 awaited kidneys.

What We Learned about the Universe in Era 209

<u>Supermassive Black Holes</u>

As we saw in Era 206 (Roaring Twenties, Prohibition, The Great Depression, World War II), through the work of astronomers Henrietta Leavitt and Edwin Hubble we learned that the Milky Way Galaxy is but one of billions of galaxies in the Universe. Also in Era 206, we saw that Karl Jansky invented the radio telescope and detected a radio source coming from the direction of the constellation Sagittarius. In 1974, Bruce Balick and Robert Brown, using the baseline interferometer at the United States National Radio Astronomy Observatory, further studied this source. They detected what they thought to be the location of a supermassive black hole at the center of the Milky Way. This object is believed to contain more than four million solar masses, and has been designated Sagittarius A* (pronounced "A star"). It is now known that supermassive black holes on the order of hundreds of thousands to billions of solar masses are found in the centers of almost all currently known massive galaxies.

<u>Investigation of Climate Change and the IPCC</u>

In 1988 the United Nations endorsed the IPCC – the Intergovernmental Panel on Climate Change. The IPCC was established by the World Meteorological Organization (WMO) and the United Nations Environment Program. The purpose of the IPCC is to give the world an objective, scientific view of climate change. The panel's mission includes investigating the political and economic impacts of climate change, as well as its risks and possible remedies.

The IPCC has become an internationally accepted authority on climate change. Its reports have the agreement of leading climate scientists and the consensus of participating governments. Former United States Vice-President Al Gore and the IPCC shared the 2007 Nobel Peace Prize. Since the early-to-mid 1990s, the IPCC has warned that human activities have been increasing the accumulation of greenhouse gases in the Earth's atmosphere, and that action is required to prevent permanent and dangerous changes in the Earth's climate system.

Space Exploration in Era 209

Pictures from Mars

The Mariner 4 mission provided the first photographs of another planet taken from space. Mariner 4 made its closest approach to Mars on July 15, 1965 and is one of the great successes of the early American space program. This mission provided valuable information that led to future missions that included landing roving vehicles on the Martian surface.

The Mariner 9 orbiter of 1971-72 yielded some fascinating discoveries. A huge canyon that dwarfs Earth's Grand Canyon was discovered and was named Valles Marineris (Latin for valley of the mariner). It is a system of canyons more than 2,500 miles long and averaging twenty miles across and five miles deep. Valles Marineris is one of the largest canyons of the Solar System. Its origins are still being explored and debated.

Voyager 2's Tour of the Outer Planets

The Voyager 2 spacecraft was launched by NASA in August of 1977. Voyager 2 was designed to visit and radio back data about Jupiter, Saturn, Uranus and Neptune, and ultimately to study the outer reaches of the Solar System.

Voyager 2 has been an exceptionally successful mission. At Jupiter, its pictures showed changes in the Jovian atmosphere since Voyager 1's visit. We learned about the volcanic eruptions on Jupiter's moon Io, which is the most volcanically active body in the Solar System. We knew that the moon Europa appeared to have streaks, and Voyager revealed these to be a collection of cracks in a thick, icy crust.

In the summer of 1989 Voyager 2 passed within three thousand miles of Neptune, the eighth planet from the Sun. Voyager 2 found six new moons, four new faint rings, and winds between 680 and 1100 kilometers/hour. Neptune's largest moon Triton is the coldest known body in the Solar System and has an "ice volcano" on its surface. Neptune certainly turned out to be far more interesting than the nondescript blue orb we knew before Voyager's flyby.

The Restless Earth in Era 209

Mount St. Helens

On Sunday May 18, 1980, Mount Saint Helens in Washington State erupted. It has been termed "the deadliest and most economically destructive volcanic event in the history of the United States." For two months prior to the actual eruption, there was a series of earthquakes and steam-venting episodes. Volcanic eruptions of this sort are typically preceded by a build-up of hot, molten rock, or magma, rising from the earth's mantle. In this instance, the build-up of the magma chamber was evident, as a huge bulge and a fracture system were visible on the mountain's north slope.

At 8:32 a.m. pacific time, the entire weakened north face of the mountain slid away due to an earthquake. The resulting landslide was the largest ever recorded, and it reduced the pressure on the molten and steam-laden rock beneath. The result was a violent eruption. Approximately fifty-seven people died as a direct result of the eruption, including persons there to study and record the event, such as geologist David A. Johnston and photographer Reid Blackburn.

An eruption column rose fifteen miles into the atmosphere and ash rained down on 11 U.S. states. A devastating effect of a volcano can be a lahar, a fast-moving mudslide caused by melting of ice fields and glaciers. In this case, the lahar traveled almost fifty miles to the Columbia River. Hundreds of square miles of land near Mt. St Helens were reduced to wasteland.

The Earth is still busily releasing the primordial heat from its formation, and is unaware of the devastating effects volcanic ash has on human machines and activity. The ash fall from the 1980 Mount St Helens eruption created major problems with water treatment systems and transportation. Many highways and roads were closed due to poor visibility. Several airports in Washington shut down for as long as two weeks due to ash accumulation and poor visibility. Fine-grained, gritty ash is not only deadly for persons unfortunate enough to inhale too much of it. It also causes major problems for internal-combustion engines and other mechanical and electrical equipment, clogging air filters and contaminating lubrication systems. It also resulted in power outages as it caused short circuits in electrical transformers. Removing and disposing of the ash was a monumental task. State and federal agencies estimated that over 2,400,000 cubic yards were removed from highways and airports in Washington.

The catastrophic events of May 18, 1980 did not deplete the rock strata beneath Mt. St Helens of all its heat and potential explosive power. Mt St Helens produced five more explosive eruptions between May and October 1980. At least 21 periods of activity occurred in the ten-year period 1980 through 1990. The volcano is still active, and shifts of magma continue, causing the re-building of the dome in its caldera.

The Earth has been a very geologically active place since its birth 4.6 billion years ago. Of course, it was far more active and tumultuous during the Hadean Era 4.6 to 3.8 billion years ago. At that time the Earth was being bombarded by meteors, asteroids and planetesimals during the early era of the Solar System's formation. It was difficult for crust to form during that era. But in our current era, except for events such as the eruption of Mt. St Helens, changes in the Earth during a human lifetime appear to be extremely slow. I believe we have a tendency to assume that the Earth was active in the past, but has slowed down to a relative "steady state" for human convenience. Events such as the 1980 eruption of Mt. St Helens serve to disavow us of that naïve and human-centered notion and to remind us of the geological truth.

Summary Comments on Era 209

(years: 1974 – 1993)

Era 209 saw major events in international affairs, and presidents with divergent styles. Gerald Ford became President when Richard Nixon resigned to avoid certain impeachment and almost certain conviction. Ford had always been a conciliator and immediately pardoned Nixon. Ford also granted a conditional pardon to those who had evaded the draft during the extremely unpopular Vietnam War. Ford considered himself an internationalist and worked with other nations to try to achieve better international cooperation.

Jimmy Carter was our 39th President, succeeding Gerald Ford. Carter established the Departments of Education and Energy, hoping the latter would stimulate development of alternatives to fossil fuels. Carter hosted the heads of state of Israel and Egypt, worked tirelessly with them, and helped create the Camp David Accord that reduced tension between those two nations. Despite that notable achievement, Carter's popularity went into a steep decline due to Americans' feelings of powerlessness when insurgents in Iran stormed our embassy and kept American citizens hostage for over a year.

Ronald Reagan's credentials included being a supporting actor, president of the screen actors' guild and governor of California. He may be the most overrated President in American history. Prior to becoming President, he backed Barry Goldwater's presidential bid in 1964 despite Goldwater's racially biased platform. As governor of California, he used police force to repress student protests. Despite the resulting deaths, he showed no remorse in later years. As President, Reagan gave enormous tax cuts to the wealthiest Americans. The stock market did well, but in his first term in office Reagan added more to the national debt than had all previous Presidents before him combined. In exchange for the debt, wealthy people purchased unprecedented numbers of large homes, luxury cars and high-end retail goods. These items have done little to defend our shores, renovate our infrastructure, or educate citizens. The Soviet Union dissolved in 1989-1991, and historians indicate that the Soviets fell apart due to their own failing economy. Some believe that it was the U.S. military build-up under Reagan that caused the Soviet Union's collapse. Reagan was popular throughout his presidency and, largely due to this, he is still rated highly by some who rate Presidents, though there is little to support such a conclusion.

In 1989 George H.W. Bush succeeded Ronald Reagan as President. Bush signed the North American Free Trade Agreement (NAFTA), an accord between the U.S., Canada and Mexico. President Bush had to grapple with serious foreign affairs issues. He launched Operation Just Cause and took military action that resulted in the restoration to office of Panama's elected president Guillermo Endara and brought drug trafficker Manuel Noriega to justice. In addition, under Saddam Hussein Iraq had invaded its neighboring country of Kuwait. Bush won overwhelming United Nations approval to undertake a multinational military response. In a military action called Operation Desert Storm, the United States and its allies defeated Hussain and Iraq and freed Kuwait. Bush pulled our forces out without capturing Saddam Hussein, and that decision may have had a profound influence on future U.S. military action under Bush's son, President George W. Bush.

Era 209 saw a steep decline in union membership in the private sector of the American economy. As previously indicated, there have been pendulum swings that have put either industry or labor in the ascendency, with more moderate or reasonable accommodations not prevailing.

The first mobile phone was developed, though affordable and widely used devices would wait until Era 210. Medical science took a major leap forward with the development of magnetic resonance imaging (MRI), a staple of today's medical diagnostic armamentarium. Also, in the medical arena, organ transplants numbered in the thousands in Era 209. Our understanding of the Universe was increased again in this era. Radio astronomy enabled us to learn that supermassive black holes reside in the center of most, if not all, massive galaxies. And the Voyager 2 mission expanded our knowledge of the outer planets Jupiter, Saurn, Uranus and Neptune. On May 18, 1980 Mount St. Helens in Washington state erupted violently, reminding us once again of the immense power contained in the Earth's primordial heat.

ERA: 210

Years: 1993 – 2021

("9/11" Twin Towers attack, Second Persian Gulf War, Climate Change, Insurrection against the US Government)

Presidents: Bill Clinton, George W. Bush, Barak Obama, Donald Trump and Joseph Biden

1992				
Bill Clinton Al Gore	Arkansas Tennessee	Democratic	370	43%
George Bush Dan Quayle	Texas Indiana	Republican	168	37%
Ross Perot James Stockdale	Texas Illinois	Independent	0	19%

1996				
Bill Clinton Al Gore	Arkansas Tennessee	Democratic	379	49%
Robert Dole Jack Kemp	Kansas California	Republican	159	41%
Ross Perot Pat Choate	Texas Texas	Independent	0	8%

William Jefferson Clinton

William "Bill" Clinton was born in 1946 and served as our 42nd president from 1993 to 2001. Bill Clinton was born in Hope, Arkansas and was named William Jefferson Blythe, Jr. His father died in a motor vehicle accident three months before the future

President was born. After his mother completed nursing school, she married Roger Clinton and Bill became William Jefferson Clinton. Clinton attended Hot Springs High School, noted for its academic rigor. In high school, Clinton was an avid reader, a student leader and a musician. He participated in Boys Nation, an American Legion program providing leadership training, and spoke about these experiences to community groups. He recalls that at age sixteen he believed he could become a musician, doctor or politician, but that the latter career choice was the only one of the three at which he believed he could excel. He identified two events that solidified his intention to be of service to others as a statesman. One was meeting President John Kennedy while a Boys State Senator on a visit to the White House. The other was hearing Martin Luther King's "I have a dream" speech, which so impressed him that he later memorized it.

Clinton graduated 4th in a senior class of 363 and went on to attend Georgetown University. In 1968, he received the degree of Bachelor of Science in Foreign Service. In 1964 and 1965, he won elections to be class president. From 1964 through 1967, he served as an intern and clerk in the office of Senator J. William Fulbright.

Clinton was an excellent student, won a Rhodes Scholarship and studied for a year at Oxford before returning to the United States in 1970 to study law at Yale. While at Yale he met Hillary Rodham, also a Yale law student. They were both idealistic and energetic, and both had a desire to put their idealism and energy to good use for their country. They married in 1975.

After law school, Clinton returned to Arkansas and became a law professor at the University of Arkansas. He ran for attorney general of Arkansas in 1976 and was elected. In 1978 he was elected governor and, at age 32, was the youngest governor in the state's history. He worked primarily on educational reform and on the state's roads. One of Clinton's advisors, Dick Morris, had surveyed voters and had concluded their number one issue was road improvements, and suggested they would pay higher license fees to obtain them. Clinton thus used a motor vehicle tax to effect road improvement, but this angered some voters and he was not re-elected. He worked at a law firm but, in 1982, he asked the people of Arkansas for a second chance and was re-elected governor. He held the position for ten years. He worked for meaningful reforms and became a leading figure among politicians known as "new democrats," a group which favored welfare reform and smaller government.

A centerpiece of Clinton's gubernatorial career was education reform. Governor Clinton called a special legislative session in 1983 and reforms were passed. They included more funding for schools, opportunities for gifted children, vocational education, higher teacher salaries, and measures to ensure teacher competency. Clinton won national recognition for having improved an educational system that had previously been considered among the nation's most deficient. While governor, Clinton chaired the National Governors' Association and the Democratic Governors' Association. In addition, he chaired the Democratic Leadership Council, a moderate group seeking to appeal to mainstream voters. Clinton ran for President and defeated incumbent President George Bush in 1992. He appealed to people as a social moderate and a fiscal conservative. He was exceptionally articulate and intelligent, and capable of quickly grasping the details and complexities of issues.

Clinton was what some people call a "policy wonk," a term referring to a person who delves into the specifics of issues and the specifics of proposed legislative remedies. In addition, his Cabinet and administrative appointments reflected the changes he tried to bring about in the nation regarding improved opportunities for women and for persons of color. He appointed more women and persons of color to positions than ever before.

The North American Free Trade Agreement, originally signed by President Bush, was enacted under Clinton. His policy of fiscal conservatism led to federal budget surpluses in 1998, 1999 and 2000. He is the last President to have balanced the federal budget. In addition, Clinton presided over the longest period of peacetime economic expansion in American history.

In August of 1993, Clinton signed the Omnibus Budget Reconciliation Act. This act cut taxes for 15 million low-income families, made tax cuts available to 90 percent of small businesses, and raised taxes on the wealthiest 1.2 percent of taxpayers. This act was of benefit to millions of individuals and small businesses. However, as it raised taxes on the wealthiest 1.2 percent, not a single Republican in Congress voted for it.

Clinton championed the Family Medical Leave Act, which to this day helps millions of workers avoid penalties at their place of employment when they must be away from work for matters related to their own health or that of their family. He also signed into law the Brady Handgun Violence Prevention Act, which established a national computer network to check the background of gun buyers, and which also required a five-day waiting period prior to the purchase of firearms.

After World War II, there were more jobs available than there were qualified persons to fill them. Employers began offering health insurance as an incentive. Americans have become accustomed to thinking that health insurance and employment are related. People working for the federal, state or municipal government, or for a large business, could usually get affordable health insurance at group rates. People who were not so affiliated could not get group health insurance coverage. Forty million Americans were without health insurance, and opinion polls revealed the public's desire for a remedy.

Despite the many successes of his administration, Clinton was unable to remedy this inequity and to reform the country's healthcare system. Republicans had decided, in advance, that they would oppose any such plan. Nonetheless a huge effort was made, in which members of every healthcare profession were brought in for lengthy consultations. The plan featured what was called "managed competition," so that it would retain the advantages of competitive free enterprise. It also had provisions to be sure that citizens who were not part of a group, such as a state, municipality or large business could still become part of a "buying cooperative" and would not be at a disadvantage when seeking the best healthcare. Despite its careful construction and obvious advantages, the plan did not win sufficient support to pass. The nation's four largest insurance companies reportedly supported the plan but Republicans opposed it. Further, persons and businesses profiting from the system in place ran misleading ads to diminish support for the plan. First Lady Hillary Rodham Clinton headed the task force that created the proposed plan, and some critics questioned whether a First Lady should have had that much influence.

On the foreign policy front, there were several military actions during Clinton's presidency. In 1996 the State Department warned about Osama bin Laden. In 1998 al-Qaeda bombed U.S. embassies in East Africa. In response, Clinton ordered several military missions to capture or kill bin Laden. Unfortunately, they were unsuccessful.

Clinton had warned Congress about the threat Saddam Hussein posed, not only to stability in the world, but to his own people. Clinton launched a four-day bombing campaign named *Operation Desert Fox* in December of 1998. Clinton stated:

> So long as Saddam remains in power, he will remain a threat to his people, his region, and the world. With our allies, we must pursue a strategy to contain him and to constrain his weapons of mass destruction program, while working toward the day Iraq has a government willing to live at peace with its people and with its neighbors.

Clinton's second term was marred by sexual impropriety. He was impeached by the House of Representatives and was tried by the Senate for perjury and obstruction of justice. He was acquitted on both charges. Despite that set of problems, the country appears to have been more concerned about the benefits Clinton's administration brought than about his sex life. He balanced the budget, shepherded through meaningful welfare reform, provided important family medical leave benefits, and provided stricter penalties for serious crimes. He presided over the longest period of stability and prosperity the nation could remember. Crime was at a twenty-five-year low and the unemployment rate was at a thirty-year low. Most of all, he showed that government can protect the rights of its citizens and can be sensitive to and responsive to the needs of the people while maintaining fiscal discipline. Clinton left office with as high an approval rating as did any modern President. Fifty-eight percent of people polled said he would be remembered as either "outstanding" or "above average" as a president, and Gallup polls in 2007 and 2011 showed that Clinton was regarded by thirteen percent of Americans as the greatest President in U.S. history. His popularity appears to have been due to his being seen as intelligent, dedicated, and as sincerely trying to improve the nation and all its citizens rather than to serve special interests or a small portion of the populace.

Bill Clinton left the White House in 2001 and moved to Chappaqua, New York, the home-base of that state's junior U.S. Senator, Hillary Clinton. In 2004 he published his memoir, *My Life*, and he oversaw the creation of his presidential library in Little Rock, Arkansas. His commitment to the betterment of society was expressed by his 2007 publication: *Giving: How Each of Us Can Change the World*.

Bill Clinton has continued to be active in public life, and has devoted himself to issues of public concern, especially through the work of the Clinton Presidential Foundation. The Foundation's agenda includes combating HIV/AIDS, fostering racial and ethnic reconciliation, and promoting the economic empowerment of poor people.

As early as 1990 the United Nations Intergovernmental Panel on Climate Change had been informing its member nations of evidence of changes to the world's climate system. In 2005, through the Clinton Foundation, Clinton championed an initiative to support programs regarding climate change. He and his predecessor, George H.W. Bush, have worked together to raise funds to help victims of natural disasters. Clinton has spoken at Democratic National Conventions in the years subsequent to his presidency.

2000				
George W. Bush Dick Cheney	Texas Wyoming	Republican	271	47.9%
Al Gore Joe Lieberman	Tennessee Connecticut	Democratic	266	48.4%

2004				
George W. Bush Dick Cheney	Texas Wyoming	Republican	286	50.7%
John Kerry John Edwards	Massachusetts North Carolina	Democratic	251	48.3%

George Walker Bush

George W. Bush was born in 1946 and served as our 43rd president from 2001 to 2009. His presidency was characterized by a terrorist attack on the country, war with Iraq, cronyism, rising federal deficits and an economic recession brought on by a real estate crash.

George W. was born on July 6, 1946 in New Haven, Connecticut. He inherited wealth, prestige and family connections. His father moved the family to Texas when young George was two years old. The family endured a tragedy when his sister died of leukemia when George was seven. George began to be a jokester and a teaser, perhaps to try to distract his mother from her grief. Bush attended prep schools, first in Texas and then in Andover, Massachusetts. He liked sports and socializing.

At age eighteen he returned to Texas to help his father run for the U.S. Senate. His interest in politics faded when his father was defeated. His father and other family members had attended Yale, and he was accepted there. Bush later admitted that he had been accepted to Yale by "legacy preference," and that he later came to believe that college admission should be based on merit. At Yale, Bush was known to be somewhat disheveled and to like fraternity life, pranks, sports, partying and flirting.

From age twenty-two through age forty Bush's life had no significant focus or direction. After college his family connections enabled him to secure a slot in pilot training for the Texas Air National Guard. There was little chance that this unit would be called into combat in Vietnam. The war in Vietnam was questionable at best, and thousands of young men, including former President Bill Clinton, used a similar stratagem to avoid combat in what they felt was not a noble cause.

The future president received preferential treatment in the National Guard. He wanted to work in the Senate campaign of Alabama Republican William Blount, and managed to arrange a transfer to a unit there. Although he claimed that he had fulfilled his National Guard duties there, no evidence exists to support that contention. General William Turnipseed would have been Bush's commander in Alabama, and he had no memory of Bush.

Bush helped with another unsuccessful Senate run by his father. At age 27 he returned to school, this time to Harvard Business School where he obtained a master's degree in business administration. He returned to Texas and started an oil business. Bush later professed to have started the business on his own and with limited funds. It appears that Bush's family contacts included wealthy investors who backed Bush's company. Making false statements about preferential treatment appears to have been a pattern for him. Despite family connections and investors, he was never a success in that field. At age thirty-one he married Laura Welch and he made another attempt at politics, this time running for Congress. He lost the election and felt that the defeat was a character-building event. But he had little direction and was not successful in business. He drank heavily, his wife threatened to leave the marriage, and his parents arranged a meeting with evangelist Billy Graham. He became somewhat of a religious zealot, and this carried through to his years as President. He became intoxicated at his own fortieth birthday celebration, and abruptly gave up drinking, smoking and chewing tobacco.

He decided that the road to success in politics might be to first attain visibility in another field of endeavor. In 1989 he assembled a group of partners and bought the Texas Rangers baseball team. The team's fortunes improved and his reputation grew. In 1994 he ran for governor of Texas against incumbent Ann Richards. Richards had vetoed a bill that would have allowed Texans to carry concealed weapons. One of Bush's campaign promises was to pursue another bill to allow concealed weapons and he was elected. He was popular as governor and was credited with improving the state's literacy rate. Claiming that the measure would reduce insurance costs, Bush persuaded the legislature to limit punitive damages that workers and consumers could obtain in suits against business. The promised reduction in insurance costs did not materialize but insurance companies did make bigger profits.

Bush won re-election as governor in 1998. He was a strong supporter of the death penalty and opposed a bill that would have replaced the death penalty with life sentences for severely retarded criminals. He sold his share of the Texas Rangers for fifteen million dollars, a handsome return on his $640,000 investment.

Bush won the Republican Party's nomination for President in 2000. He defeated John McCain in the primaries and faced Al

Gore in the general election. Many voters disliked Bush's tenuous grasp of issues, his inexperience and his poor command of the English language. *Time Magazine* called his experience "one of the thinnest resumes in a century." Consumer advocate and Green Party candidate Ralph Nader drew enough Democratic votes from Al Gore to make the difference. Bush lost the popular vote to Al Gore, but won the electoral college, 271 to 266. The state of Florida, of which his brother Jeb was governor, voted for Bush by five hundred votes, but there were charges that Gore had been cheated out of thousands of votes. The matter went to the Supreme Court, which voted five-to-four against overturning the election results. By the vote of one justice, Bush became President.

Bush's predecessor, Bill Clinton, had managed a budget surplus for three consecutive years. Upon taking office Bush's budget office estimated there would be a 5.6 trillion-dollar surplus over ten years, but this was not to be. During Bush's administration, federal spending increased sixty percent from 1.8 trillion to 3.0 trillion dollars while revenues increased by only twenty-five percent from two trillion to 2.5 trillion. This was due in part to the largest tax cut in American history that Bush championed. The last three years of President Clinton's administration saw budget surpluses. But the trend of very fast increases in the federal budget deficit, begun in Ronald Reagan's first term, were re-started under Bush's administration.

Hurricane Katrina devastated the Gulf Coast, and particularly the city of New Orleans, in late August, 2005. The Federal Emergency Management Administration (FEMA) appeared to do a very poor job of helping New Orleans and its citizens. Some observers contend that President Bush was given to making important appointments based on "cronyism" rather than on qualifications. In this instance Michael Brown, former president of the Arabian Horses Foundation, had been Bush's pick to head FEMA. Brown resigned on September 12, 2005 following his controversial handling of the Hurricane Katrina emergency.

Michael Brown was not the only appointment by Bush that gave authority to undeserving persons. Bush appointed Jeffrey Acosta to the National Labor Relations Board. This appointment led to Acosta's appointment as U.S. Attorney for Southern Florida. Acosta gave a sweetheart deal to sex offender Jeffrey Epstein. This deal was later ruled illegal by a federal judge. Many victims were denied justice for over a decade due to this deal. In addition, in 2004 Bush appointed Matthew Whitaker as U.S. Attorney in the southern district of Iowa, even though Whitaker had no prosecutorial experience. Whitaker later served on the board of World Patent Marketing, an invention promotion company that was shut down by federal regulators due to deceptive marketing practices and scamming consumers out of twenty-six million dollars. Much damage has been done by Bush's habit of giving appointments to undeserving friends.

Bush's presidency was colored mostly by foreign policy issues, notably the war in Iraq. On September 11, 2001 the twin towers of the World Trade Center in New York collapsed after suffering direct hits from two planes, which ignited devastating fireballs caused by the impact and fuel, while a simultaneous explosion occurred at the Pentagon in Washington. There are some irregularities regarding these events that are far beyond the scope of this book. Bush initiated military action against the Taliban in Afghanistan. He routed the Talban and installed a new government. However, the organization known as al-Qaeda survived, and its leader Osama bin Laden escaped.

Bush proposed and signed a new law, the Uniting and Strengthening America by Providing Appropriate Tools Required to intercept and Obstruct Terrorism Act of 2001 (the PATRIOT Act). This expanded the government's ability to find and arrest people. It was, of course, aimed at terrorists, but critics contend that the act increased the government's ability to invade the privacy of all citizens and not just terrorists.

By March of 2003 Bush turned his attention to the task of overthrowing Iraq's dictator, Saddam Hussein. Bush contended that Hussein possessed weapons of mass destruction, though no clear evidence of such weaponry existed, either at that time or after the war. Bush also contended that Iraq was aiding or harboring al-Qaeda, but that was also never confirmed. There is more than one interpretation of Bush's decision to declare war on Iraq. Some observers believe that Bush wanted revenge on Hussein for an attempted assassination of his father, President George H.W. Bush. Others contend that he felt that the elder Bush had withdrawn from Iraq too soon in the previous war, as he had left Hussein in power, and that the younger Bush therefore wanted to complete his father's unfinished job by removing Hussein.

It is important to note that Iraq is not truly a unified country. In 1921, a British-backed monarchy linked various factions together. There are various factions within Iraq, notably Kurds, Shiites and Sunnis, and these factions have very strong conflicts among them based on religious affiliations and tribalism. Thus, the goal of establishing a government in Iraq with the ability to satisfy most of its citizens and maintain stability is probably unrealistic.

Regardless of his intentions, Bush ordered an invasion of Iraq. There is no question that Saddam Hussein was a brutal and dangerous dictator, under whose rule terrible atrocities against his own citizens were carried out, encouraged or condoned. However, Bush was unable to assemble the broad array of supportive nations that his father had managed in the 1991 Iraq war. Although the global community had been supportive of our military action in Afghanistan, many opposed the war in Iraq. The dissenters included two of our closest allies, France and Germany. Americans were divided as to supporting an invasion of Iraq.

Bush assembled what he termed the "coalition of the willing," which included the U.S., Great Britain and a few smaller nations. The invasion was quickly successful. Saddam Hussein went into hiding and was captured by U.S. troops in December 2003.

Hussein was executed by the Iraqis on November 5, 2006. Americans appointed a ruling council and more than 150,000 American troops remained in Iraq. The reasons for having invaded Iraq were not validated. No weapons of mass destruction were ever found and the suspicion about Iraqi ties to al-Qaeda were not confirmed.

The invasion of Iraq ushered in a very frustrating time for American policy in Iraq, and for economic and political life here at home. Violence and civil war continued in Iraq and many American soldiers lost their lives. The military expenditures were immense. The United States has typically presented itself as humanitarian and as a defender of decency and human rights. That image was dealt a terrible blow when it was discovered that there was considerable abuse and torture carried out near Baghdad at the Abu Ghraib prison. The roster of detainees and the treatment they received was hidden from the Red Cross, in violation of international law and military policy. The *New York Times* reported, "The newly disclosed documents are the latest to show that such activities were known to a wide circle of government officials." The international image of the United States was severely tarnished. The American economy suffered, the price of oil spiked, and Bush's approval ratings plummeted. Further, there were scandals within the Republican Party.

In addition, there was a bubble in the real estate market, brought about by greedy and unscrupulous interests. Financial institutions were lending money far too easily. They were giving what have been termed "NINJA" loans, an acronym for "no income, job or assets." They were able to issue such loans because of collusion with insurance companies. The mortgage originators would bundle the loans together and sell them as "securities," a type of bond. There is nothing unusual about bundling mortgage loans together and selling them as bonds. If the loans were made to credit-worthy individuals, they may be good investments. But in this case, the loans were made to spectacularly non-credit worthy persons, and the practice was enabled by insurance companies who gave these loans high ratings, although they were clearly, in the parlance of the industry, "junk bonds." When the loans were not repaid, the holders of the "securities" asked the insurance companies to repay them, and there were no reserves from which to pay. It had been a Ponzi scheme. The real estate market, the stock market and the entire world economy went into decline. Whether Bush's administration should have been aware of the situation, and whether steps could have been taken by the government to avert it, is open to question. But it did occur on Bush's watch and it did mar his presidency.

Bush's presidency appears to be a record of modest successes and spectacular failures. He took powers to himself that may have upset the balance among our three branches of government. And he did leave the country more divided and insecure than he had found it.

George W. Bush left office with a 33 percent approval rating and with 60 percent of the American public believing history would consider his presidency below average. Bush returned to Texas and briefly stayed at his Prairie Chapel Ranch in Crawford. He seemed to enjoy the relaxation and time away from power.

George and Laura bought a home in an exclusive Dallas neighborhood near Southern Methodist University and became a part of the Dallas community. He hosted barbecues at his home and supplemented his income with paid speeches. He established the George W. Bush Institute at SMU. In 2010, he published *Decision Points*, describing his experiences and policies as President. In his free time he enjoyed golf, Texas Rangers baseball games, and reading American history.

Bush also engaged himself in charity work. He sponsored golf tournaments and an annual mountain bike ride to raise funds to help wounded veterans. Bush also began painting, portraying his pets and world leaders whom he had met. His paintings attracted national attention and were displayed in his presidential library.

2008				
Barack Obama Joe Biden	Illinois Delaware	Democratic	365	52.9%
John McCain Sarah Palin	Arizona Alaska	Republican	173	45.7%

2012				
Barack Obama Joe Biden	Illinois Delaware	Democratic	332	51.1%
Mitt Romney Paul Ryan	Massachusetts Wisconsin	Republican	206	47.2%

2016				
Donald Trump Mike Pence	New York Indiana	Republican	304	46.7%
Hillary Clinton Tim Kaine	New York Virginia	Democratic	227	48.9%
Gary Johnson William Weld	New Mexico Massachusetts	Libertarian		3.3%
Jill Stein Ajama Baraka	Massachusetts Washington DC	Green		1.1%

Barack H. Obama

Barack Obama was born in 1961 and served as our 44th president from 2009 to 2017. Obama was born on August 4, 1961 in Honolulu, Hawaii, two years after Hawaii became the 50th state. Obama's father was black and his mother was white. His father, Barak Obama, Sr., began life as a goatherd in Kenya. He was a gifted student and was among those chosen by Kenyan leaders and American sponsors to study in the United States. In his three years at the University of Hawaii, he gained a reputation as a brilliant and charismatic man with an eagerness to help his native developing nation. Barak's mother, Stanley Ann Dunham, born in Kansas, graduated from a Seattle, Washington high school and was remembered by classmates as intellectually mature.

The future President's parents met at the University of Hawaii in 1960 and married in 1961. The senior Obama accepted a scholarship for post-graduate work at Harvard, a scholarship with no provisions for re-locating a family. The stress of separation led to eventual divorce in 1964. Barak's mother continued her studies, married another international student in 1967, and moved with young Barak to Indonesia. Obama's experience in Indonesia contained extremes. He experienced the pleasures of being in a Pacific Island nation, with crocodiles and birds of paradise in the backyard. But he also witnessed crop failures, floods and poverty. Seeing many extremely impoverished persons gave him an attitude of both realism and compassion.

Barak was to be the first person of color to be elected President of the United States. At age ten Barak returned to Hawaii to live with his grandparents and to attend a prestigious school while his mother was in graduate school. Reflecting on his years in Hawaii, Obama later wrote, "The opportunity that Hawaii offered — to experience a variety of cultures in a climate of mutual respect — became an integral part of my world view, and a basis for the values that I hold most dear."

From an early age Barak experienced the occasional alienation brought on by his bi-racial heritage, and he struggled to understand the world from the perspective of a black American. As a high school student, he had a passion for basketball, and was a key player on a team that won the Hawaii state championship. He attended Occidental College near Los Angeles, where he became involved with student activism. He transferred to Columbia University in New York and graduated in 1983 with a bachelor's degree in political science and international relations.

Obama then worked in New York for a public interest research group and for a consulting firm. Working for community change appealed to him and at age 24 he accepted a position in Chicago. He successfully developed programs in tutoring, tenants' rights, job training and employment advocacy. He attended Harvard Law School and graduated magna cum laude. He attracted national attention as the first black president of the Harvard Law Review. After graduating from Harvard, he accepted an offer to be a visiting fellow at the University of Chicago Law School, where he worked on his first book. He taught at the University's Law School for twelve years, specializing in constitutional law, due process, equal protection and voting rights.

In 1992 Barak married Michelle Robinson, also a lawyer. The Obamas' two daughters were born in 1995 and 2001. In 1995 Obama was elected to the Illinois state senate, where he helped pass ethics legislation and a law designed to restrict racial profiling. Years later he looked back on his seven years on the state senate and said,

> I learned that if you're willing to listen to people, it's possible to bridge a lot of the differences that dominate the national political debate . . . I pretty quickly got to form relationships with Republicans, with individuals from rural parts of the state, and we had a lot in common.

Obama gained recognition when he gave the keynote speech in the 2004 Democratic National Convention, and he won a seat in the U.S. Senate in 2005. While in the Senate, he served on a half-dozen committees, including Foreign Relations, Veterans' Affairs and Homeland Security & Governmental Affairs. When he announced his candidacy for the presidency, he emphasized the goals of ending the Iraq war, energy independence and reforming the healthcare system. In June of 2008, his opponent Hillary Clinton withdrew from the presidential race and endorsed Obama and, in August, he selected Delaware Senator Joe Biden to be his running mate. The Obama/Biden ticket defeated the John McCain/Sarah Palin ticket and Obama became the first African American to become President.

Obama inherited a difficult economic situation. After the bank debacle of 2007-2008, former President Bush had provided a huge bailout for the banks. There was divided opinion about Bush's move. Some observers felt the banks had created their own problem and should have been allowed to fail. Others contended that failed major banks would have destabilized the world financial picture as well as the United States' place in it. This author has insufficient knowledge of international finance to have an opinion on that matter. What is clear is that Bush had set us on a course and that Obama had little choice but to follow it. Obama's team worked closely with the outgoing administration of George W. Bush, and an economic stimulus package was Obama's first order of business.

The expense of the war in Iraq and the Bush tax cuts had greatly increased the national debt, which was then further increased by the enormous bank bailout. There was a widely held opinion that to fail to provide a monetary stimulus to the economy would bring about immediate recession. The best hope appeared to be to continue to provide monetary stimulus (euphemistically called "quantitative easing"), and to hope that the economy would eventually grow sufficiently for revenues to catch up with expenditures and debt. Thus, Obama signed the Economic Recovery and Reinvestment Act of 2009.

The United States had been the only major industrialized country in the world without universal healthcare, which many felt was a national embarrassment. What most people do not realize is that the health insurance situation at that time was an historical accident. Careful thought leads to the conclusion that employment and healthcare have nothing to do with one another. After World War II, however, there was a "sellers' market" for jobs. (There were not enough qualified persons to meet the needs of business and industry). In order to attract qualified job seekers, businesses and corporations offered health insurance as an incentive. People eventually became accustomed to this idea and, as a nation, we went along with it without further analysis. Millions of people were without health insurance because they did not work for the federal government, or for a corporation, city, state, or municipality that offered it as an employment benefit. Purchasing it individually without the buying power of a large group was prohibitively expensive.

Finally, in March of 2010, the Patient Protection and Affordable Care Act was passed and Obama signed it into law. This was a huge accomplishment and provided health insurance for millions who had been without it.

Wars in Iraq and Afghanistan continued to be a thorn in the side of American foreign policy. Obama contended that Afghanistan was the more important war in terms of fighting terrorism. Obama proposed to send additional troops to Afghanistan to more forcefully conduct the war. The plan was to withdraw troops after eighteen months and to turn over full responsibility for that country's security to the Afghans.

By August of 2010, it appeared Obama would accomplish his stated goal of withdrawing all U.S. troops from Iraq. However, when the militant group ISIS gained ground, captured the city of Mosul, and embarked on a campaign of massacres and ethnic cleansing, there appeared to be no alternative to assisting Iraq's beleaguered security forces. Eventually Obama's goal was reached and, by 2012, no American troops were fighting in Iraq.

By 2010, CIA operatives discovered Osama bin Laden living in a compound in suburban Pakistan. Obama rejected a plan to bomb the compound and instead authorized Navy SEALs to conduct a "surgical raid." The operation took place on May 1, 2011. The result was the death of bin Laden and seizure of information. Bin Laden's identity was confirmed by DNA testing and he was immediately buried at sea. News of bin Laden's death was received positively at home and overseas.

In addition to his admired oratorical skills, Barack Obama is an accomplished author. Many of his books have become bestsellers. He consistently provides an image of devotion to democratic ideals, tolerance and a reasoned approach to issues. Though reasoned, his communication style and message are inspirational, and he calls for the best in us. He is a voice in favor of human rights and environmental protection, and he provided the United States with a solid boost in terms of its overseas reputation. Polls in Europe indicated that Obama was regarded as the most respected world leader. In 2018, the American Political Science Association surveyed historians, who ranked Obama the eighth-greatest American President.

In his final press conference as President, Obama indicated a hope to avoid further active involvement in politics. However, he stated he would reverse that intention under certain circumstances, notably the following:

*** systematic discrimination being ratified in some way,

*** the creation of obstacles to voting,

*** efforts to silence the press or dissent.

(All three conditions would occur within four years).

The incoming Trump administration seemed determined to eradicate Obama's achievements in the realms of climate change, immigration, health care, and financial regulation. Although he typically refrained from criticizing the new administration, Obama took issue with some of Trump's policies and the direction in which Trump was taking the country. Specifically, Obama was critical of Trump's decision to withdraw from the Paris Agreement on climate change. Also, in 2015, there had been a deal, known as the Joint Comprehensive Plan of Action, but typically referred to as "The Iran Nuclear Deal." Iran had agreed to dismantle reactors and halt nuclear-related programs in exchange for the United States and other countries to lift economic sanctions. Trump withdrew from that agreement, and Obama was critical of that decision.

The Obama Foundation was created in January 2014 and Jackson Park, on Chicago's South Side, was chosen as the location for the Obama Presidential Center. The Center's mission included inspiring future leaders as well as serving as an economic engine for the South Side. The Center was designed to include a library, museum, athletic facility, and forum for public meetings. It would also serve as the headquarters for the Obama Foundation and for an organization Obama founded in 2014 to provide opportunities for boys and young men of color.

For at least three decades Republicans had been persistently redrawing voting districts in an unnatural way to favor Republican candidates. That process is known as "gerrymandering." The National Democratic Redistricting Committee, an organization led by former attorney general Eric Holder, is focused on countering the abuses of Republican gerrymandering. Former President Obama indicated his support for Eric Holder's organization.

Shortly after Joe Biden's successful election in 2020, Obama released the memoir, *A Promised Land.* It is the first of two proposed volumes, and describes his early life through the events of May 2011. The documentary TV series *Obama: In Pursuit of a More Perfect Union* was released in 2021.

Donald Trump

For several reasons, this president is the most difficult about whom to write. Events were unfolding as these words were being written. Thus, this writer has been engulfed with day-to-day information – enough to fill many volumes.

Background

Donald Trump is the third in the line of entrepreneurial men. His grandfather, Friedrich Trump, began the family's real estate

309

business, and had a reputation for illegal business practices. Donald Trump's father, Fred C. Trump, continued to build the real estate business. Fred Trump also had the reputation of being an unscrupulous businessman, as well as being a virulent racist and Ku Klux Clan sympathizer.

A New York Times's investigation, based on a vast trove of confidential tax returns and financial records, reveals that Donald Trump received today's equivalent of at least $413 million from his father's real estate empire. The Times' investigation of the Trump family's finances is unprecedented in scope and precision. The findings are based on interviews with Fred Trump's former employees and advisers and more than 100,000 pages of documents describing the inner workings of his empire. They include documents culled from public sources: mortgages and deeds, probate records, financial disclosure reports, regulatory records and civil court files.

Donald Trump helped his parents evade taxes, resulting in a much larger inheritance. Records and interviews revealed that Trump and his siblings set up a sham corporation to disguise millions of dollars in gifts from their parents. Donald Trump helped his father take improper tax deductions and helped formulate a strategy to undervalue his parents' real estate holdings by hundreds of millions of dollars to reduce estate taxes.

Trump has been sued hundreds of times, mostly for failing to pay people for their work. He also refused repayment to New York banks, who finally refused to lend to him. Trump went overseas and for years received hundreds of millions of dollars in unsecured loans from Deutsche Bank. Deutsche Bank has ties to Russia and was fined ten million dollars for money laundering schemes with Russians. This is significant as it has been a Russians practice to compromise persons of influence in other countries, and they appear to have purposefully compromised a man who, with their help, would be President of the United States.

The Russia Connection

Russian interests, specifically Russian President Vladimir Putin, were energetically involved in the U.S. 2016 presidential campaign. Russian operatives mounted an extensive Internet campaign using automated systems, or "bots," to spread negative information about Trump's opponent, former Senator and Secretary of State Hillary Clinton. Russians were also involved in an organization called "Wikileaks," which disseminated supposedly damaging information about candidate Clinton, often at very strategic moments in the campaign. Trump carried out his campaign through orchestrated, televised rallies. Several times during his rallies Trump openly encouraged the Russians and Wikileaks to continue to disseminate information.

During his campaign for the presidency, Donald Trump repeatedly denied that he had any business dealings with Russians. It was later definitively shown that, for at least six months into the campaign, Trump had in fact been negotiating with the Kremlin and high-ranking Russians for the right to build a Trump Tower in Moscow. As part of those negotiations, Trump offered Russian president Vladimir Putin a fifty-million-dollar penthouse condominium in the proposed Trump Tower.

In January of 2016, during the primary campaign, there was a meeting in Trump Tower attended by members of Trump's family and campaign organization, and by high-ranking Russians. This meeting was the subject of investigation, and evidence indicated that a primary purpose of the meeting was to help Trump's campaign for the presidency by arranging for negative press for his opponent, Hillary Clinton.

In addition, during the Republican national convention, the republicans inexplicably changed their party's platform, lessening sanctions against Russia – sanctions that had been imposed due to Russia having invaded and annexed the Crimea in 2014. This immediately followed contacts between the Trump campaign and the Russians. Twelve Russian operatives were later indicted by special counsel Robert Mueller for illegal interference with the 2016 U.S. presidential election.

Donald Trump's presidency was for many months characterized by investigations of the assistance his campaign received from Russian entities, and of Trump's persistent efforts to obstruct justice in its pursuit of the facts of this interference. On May 17, 2017, Robert Mueller was appointed special counsel to investigate these matters. The investigation was active until March of 2019. Mueller's report asserted that Trump had obstructed justice. However, based on a previous memorandum in the Department of Justice, Mueller concluded that a sitting President could not be indicted.

Donald Trump's hand-picked Attorney General, William Barr, did all he could to prevent Congress from seeing Special Counsel Robert Mueller's entire report. He therefore blocked Congress from exercising its constitutional "checks and balances" power to rectify what many observers have termed "unprecedented attacks" on the justice system, on the Constitution, and on the rule of law, committed by the Trump administration. We do know that Mueller's report found that Trump had welcomed Russian interference in the election, and that he had obstructed justice.

Tax Advantage for Wealthy

Trump managed to encourage Congress to pass one major piece of legislation - a tax reform bill that provided modest relief for average Americans but that gave enormous tax breaks for the wealthiest Americans. As an example, there is a Las Vegas casino owner who gave thirty million dollars to the Republican party after saving six hundred million dollars due to the new tax law.

Previous administrations have worked cooperatively with other nations and have had notable achievements. These include the North American Free Trade Agreement (NAFTA), The Paris Accord, in which 175 parties agreed to mitigate climate change by reducing carbon emissions, and the Iran nuclear deal that was signed by Iran and world powers to limit Iran's nuclear capabilities. Without congressional approval or the approval of most of his advisors, Donald Trump withdrew the United States from all three of those arrangements.

Immigration from Latin American countries represents a complex set of issues and has been at issue during and after the campaign. During Trump's administration, laws allowing people to legally seek asylum in the U.S. were ignored. Families with children, attempting to enter the U.S., have been separated from one another, denied legal assistance and subjected to poor conditions.

A further example of Trump's attempts to undo the work of previous administrations has been his position on the World Trade Organization – WTO. The WTO sets and implements trade rules for 164 member countries. A member country may file a complaint against another. The panel makes a ruling, which can be appealed to an appellate court, which comes to a final and binding decision. Trump told his advisors that he wants to unravel the WTO, stating that the U.S. loses most of its cases. His advisors told Trump that the U.S. in fact wins over seventy percent of its cases, and offered to show him the data. Trump refused to look at the evidence and held onto his faulty opinion.

Trump has in many ways expressed and encouraged anger and hatred. One of many examples is his response to the "Unite to Right" rally, a white supremacist rally that occurred August 11-12, 2017 in Charlottesville, North Carolina. There were riots at the rally and Trump stated, "There are very fine people on both sides." The press referred to this as "the moral equivalence issue," indicating that Trump was stating that promoting acceptance on the one hand, and promoting hatred and discrimination on the other hand, are of equal merit.

Trump persistently undermined public confidence in investigative newspaper reporting. Hundreds of times he used the term "fake news" to characterize news items not favorable to him. In addition, he often characterized the media as "the enemy of the people." His use of that phrase is reminiscent of its use by dictators such as Joseph Stalin and Mao Zedong.

Donald Trump has had tense relations with heads of state of democratic nations such as Canada, Great Britain and France. At the same time, he has shown clear and obvious favoritism to known dictators such as Russia's Vladimir Putin, North Korea's Kim Jong Un, and Saudi Arabia's Mohammed bin Salman.

The people of North Korea are known to be the most oppressed and least informed people in the world. Kim Jong Un is known to be a ruthless dictator who regularly imprisons and executes his opponents, and Trump has been informed of this by the State Department. Prior Presidents would not meet with Kim Jong Un, not wanting to legitimize him. Trump agreed to meet with Kim Jong Un and did so on February 27-28, 2018 in North Korea. Trump took Jon Un's word that he had not engaged in atrocities, and has referred to Jon Un as "my friend."   After that meeting with Jon Un, Trump did something that Russian President Vladimir Putin had been wanting.   Without consulting our military, and without getting anything for the U.S. in return, Trump announced that the U.S. would discontinue joint military exercises with our ally, South Korea. The joint military exercises had the purpose of signaling North Korea (and Russia) that the United States would support the democracy of South Korea, deterring aggression against that country.   Trump's action essentially signaled his departure from the United States' seventy-year policy of containing communism. Trump's national security advisor, John Bolton, said he did not believe the Trump administration really meant its pledge to halt North Korean nuclear efforts that could threaten the United States. Bolton stated the president's outreach to Mr. Kim Jon Un had benefited only Jon Un. Bolton further opined that Trump's only concern during that meeting had been the number of journalists in attendance.

In July of 2018 Trump met with Vladimir Putin in Helsinki. Trump insisted that no other American could be present and he sternly warned the translator to never reveal what had been discussed. Trump had been informed by several US intelligence agencies that the Russians had interfered with the 2016 presidential election. In a rebuke of the US intelligence community, Trump disagreed with our own government's assessment. Trump announced publicly that he had asked Putin if he had interfered. He said that Putin vigorously denied it and he believed him, thereby siding with a dictator instead of our own intelligence community.

Journalist Jamal Khashoggi was a Saudi Arabian dissident and a columnist for the Washington Post. On October 2, 2018 Khashoggi was murdered at the Saudi consulate in Istanbul, Turkey. The Turkish government claimed there was evidence that Khashoggi was murdered by agents of the Saudi Arabian government. Saudi Crown Prince Mohammed bin Salmon stated that he bore responsibility for Khashoggi's death, but denied that he had ordered the murder. A United Nations investigation reported there was "credible evidence" that bin Salmon was responsible for the assassination of Khashoggi. Trump later stated that he asked bin Salmon if he had ordered the murder of Khashoggi. Trump stated that bin Salmon "strongly denied it" and Trump accepted that denial at face value. This is but one of several occasions on which Trump displayed his habit of accepting or denying the word of others based not on a rational assessment of evidence, but rather on whether he felt it served his purpose at that moment. It is also one of several instances in which Trump embraced dictators, while at the same time exhibiting contentious relations with leaders of democratic nations.

Trump displayed further actions that directly benefited Russian President Putin. He withdrew U.S. military troops from Syria, thereby tipping the scales of power there to the Russians. Trump's Secretary of Defense General James Mattis attempted to change Trump's mind about this move. When he failed to change Trump's mind, General Mattis resigned on December 20, 2018, stating that Trump's actions were more to the advantage of our enemies than our allies.

In August 2019 the Pentagon released a report describing Russian interference against many nations. The report indicated that the Russian attacks were particularly effective against countries that were deeply divided and that lacked the will to counteract Russian interference. During Trump's presidency, the U.S. met these two criteria.

Donald Trump used the power of his office to attempt to strongarm the government of Ukraine into announcing that they were investigating Trump's potential election rival, Joseph Biden. Ukraine needed money for military support to protect itself from Vladimir Putin's Russia. Although there were no specific allegations, and no evidence of anything to investigate, Trump insisted he would withhold that military aid unless Ukraine consented to announce an investigation as he demanded. Several members of the administration testified about these events to the House of Representatives, and Donald Trump was impeached by the House for conspiring with a foreign power to influence elections and for contempt of Congress.

Trump's National Security Advisor John Bolton later revealed that, in a critical meeting on Aug. 20 he, Defense Secretary Mark Esper and Secretary of State Mike Pompeo all tried several times to persuade Trump to release the aid to Ukraine. Bolton reported that Trump refused to release the aid until Ukraine agreed to state it was investigating Hillary Clinton and Biden.

In May and June of 2020, there were demonstrations across the country to protest the killing of a black man by a police officer in Minneapolis. In response to Trump's responses to this situation, General Mattis wrote a lengthy response which was reported in *The Atlantic*. Included in his remarks, Mattis stated,

> Donald Trump is the first president in my lifetime who does not try to unite the American people - does not even pretend to try. Instead, he tries to divide us. We are witnessing the consequences of three years without mature leadership.

During his presidency, Donald Trump has used unusually questionable means to secure his reelection. John Bolton reported that Trump asked Chinese President Xi Jinping to help him win reelection, informing Xi that increased agricultural purchases by Beijing from American farmers would aid his campaign.

As the 2020 election approached, Trump's strategies to secure reelection became more frantic and more illegitimate. As he believed mail-in ballots would help his opponent, Trump expressed strong and unwarranted skepticism about mail-in election ballots. He appointed a new postmaster general for the apparent purpose of slowing mail delivery. The new postmaster general, known to be a Republican "megadonor," took mail processing equipment off line, eliminated overtime for postal workers, and reduced the post office's ability to process mail-in votes. Two weeks before the election, Trump tried a strategy more typical of dictators than of democratic leaders. He demanded that the Attorney General open an investigation of his opponent for the presidency, Joe Biden. Trump did not indicate the alleged crime to be investigated.

The U.S. Senate conducted what was supposed to be a trial relative to Trump's impeachment. However, the Senate voted not to hear testimony from witnesses, a very unusual method to conduct a trial. Furthermore, all but one Republican senator voted against Trump's removal. Trump's former National Security Advisor John Bolton later blamed House Democrats for what he termed "impeachment malpractice." First, he stated that they should have waited for the court system to rule on whether witnesses like him should testify. Secondly, he faulted them for restricting their inquiry solely to the Ukraine matter. He felt there were many more examples of misconduct by the president that could have been used to remove Trump from office.

On November 3, 2020 a record number of American voters – more than 155 million - went to the polls for our national election. Trump received strong support but Joseph Biden received seven million more votes. On November 7 several major media outlets declared Biden the winner, and therefore President-elect.

Overall, Trump's presidency was characterized by the following themes:

1) Successful efforts to immensely increase the wealth of already wealthy persons at the expense of the economy and the federal budget.

2) Repeated and persistent efforts to undermine the public's confidence in legitimate news media.

3) Creating an atmosphere in which anger and hatred are legitimized.

4) Deterioration in the quality and appropriateness of Public Comments and Discourse.

5) Clear and obvious favoring of dictators at the same time as having contentious relations with leaders of democratic nations.

6) Persistent disregard of truth.

7) Reliance on uninformed opinions and a "shoot-from-the-hip" approach instead of reliance on science, information and reason.

8) An unhealthy Relationship with Russian President Vladimir Putin, suggestive that Putin wields strong and improper influence over the former U.S. President.

9) A response to the Covid-19 pandemic characterized by neglect of leadership, denial of scientific evidence, repeated false statements, and railing out randomly against others.

10) Frantic and illegitimate efforts to remain in office, including unsuccessful lawsuits to overthrow his election loss, and inciting a violent insurrection against the United States government.

Stephen Schmidt is an American communications and public affairs strategist who has worked on Republican political campaigns, including those of President George W. Bush, California Governor Arnold Schwarzenegger, and Arizona Senator John McCain.

In June 2018, Schmidt had renounced the Republican Party. Several lifelong Republican strategists had joined Schmidt and formed The Lincoln Project, an effort to prevent Trump's reelection. On November 7, Schmidt stated,

> The American people have fired the most corrupt, the most indecent, the most incompetent president in American history. Trump is going to occupy a shameful place in American history.

Trump's shameful place in history has become much worse since Stephen Schmidt uttered those words. Despite Joe Biden's victory in the election, Trump would initiate a controversy resulting in an insurrection against the United States government, threatening the continuation of the American experiment in democracy.

Trump did not concede the election and, without any supporting evidence, claimed that the states' vote counts were fraudulent and that the election had been stolen from him. In the weeks after the election, Trump's lawyers filed sixty lawsuits in several states, including Georgia, Pennsylvania, Michigan and Arizona. The lawsuits sought to overturn the election results due to fraud but, as no evidence of fraud was presented, all these lawsuits were dismissed.

On November 16, 2018 Congress had passed the Cybersecurity and Infrastructure Security Agency Act. This act established a division of the U.S. Department of Homeland Security entitled the Cybersecurity and Infrastructure Security Agency (CISA). President Trump appointed lifelong republican Christopher Krebs, a former Microsoft executive, to head CISA. Krebs coordinated federal and state efforts to strengthen election security, and ensure confidence in the integrity of the voting process.

When President Trump claimed that the election had been fraudulent, CISA Director Krebs' agency rejected the claims as "unfounded." Krebs insisted, "The November 3rd election was the most secure in American history." Trump immediately fired Director Krebs.

Having failed via sixty lawsuits to overturn the election result, Trump turned to other methods. First, he asked the military to seize voting machines in swing states such as Michigan and Pennsylvania. When the military refused, Trump made the same request of the Department of Homeland Security. When they refused, he asked his hand-picked attorney general William Barr to have the justice department seize the voting machines. Even Barr, who had been noted by many observers to have acted more as Trump's personal attorney than as Attorney General, refused Trump's request. Trump then turned his attention to Vice President Mike Pence who, as Vice President, presides over the Senate. He asked Pence to refuse to certify the election results. Pence refused this request.

Trump was not done attempting to overturn the election. During his presidency, several fringe groups of white supremacists and neo-Nazis had emerged throughout the country. It is important to note that Donald Trump did not create these people or their hatred. They had been there all along, but had remained

320

essentially dormant. These groups emerged from obscurity, having frequently been encouraged by Trump rhetoric. Trump used social media platforms Twitter, Facebook and Instagram, and implored these groups to come to Washington DC on January 6, 2021, the day the joint session of Congress was to certify the election results, making Joseph Biden President. In his social media messages, Trump said "it will be wild." On January 6, Trump addressed a large group of his supporters who had gathered. He told them to fight to keep him President. He told them to march on the capitol, encouraging them to "fight like hell," and to "show no weakness."

Trump's supporters did march on the Capitol. Some insurrectionists had a noose and shouted their intention to hang Vice President Michael Pence for his refusal to obey Trump and stop the election certification. The scene was reminiscent of events in unstable countries. Trump's supporters stormed the Capitol, causing security to rush senators and representatives to safety. The scene was violent and several people died. Trump's parting words to the insurrectionists via social media included "Go home. I love you. I know how you feel." The social media platforms Twitter, Facebook and Instagram, having for more than four years been a vehicle for Trump's consistently inflammatory messages, cancelled his accounts, stating that though they value free exchange of ideas, they cannot be used to incite violence.

The insurrection of January 6, 2021, incited by Donald Trump, was one of the darkest days in the history of our democratic republic. In his last days in office, Trump orchestrated an event that raised doubts about our ability to maintain civil discourse, the safety of elected officials, free elections and democracy.

Americans are more divided than they have been in decades – perhaps more divided than ever. The Trump administration consistently displayed a lack of knowledge of, or commitment to, the rule of law and the Constitution. Many Americans are oblivious to these developments. The two major political parties appear to have no common ground on which to stand to make compromises and to govern.

2020				
Joseph Biden Kamala Harris	Delaware California	Democratic	306	51.3%
Donald Trump Mike Pence	New York Indiana	Republican	232	46.9%

Joseph Biden

Joseph Biden was born in 1942 and is currently serving as our 46th President. Biden was born in Scranton, Pennsylvania. His family moved to Castle County, Delaware when Joseph was ten. He attended the University of Delaware and later earned his law degree at Syracuse University in 1968.

Biden began his political career in 1970 when he was elected to the New Castle County Council. He was elected as a Democrat to the United States Senate from Delaware in 1972, making him the sixth-youngest senator in U.S. history. Biden represented Delaware in the US Senate from 1973 to 2009, when he became Barak Obama's Vice President. He had been the chair or ranking member of the Senate Foreign Relations Committee for 12 years and was thus well-prepared to be influential in foreign affairs while serving as vice president. Biden was instrumental in the 2011 withdrawal of U.S. troops from Iraq.

Biden also chaired the Senate Judiciary Committee from 1987 to 1995. In that role he dealt with such issues as drug policy, crime prevention, and civil liberties issues. He championed the Violent Crime Control Act, the Law Enforcement Act and the Violence Against Women Act. He oversaw six U.S. Supreme Court confirmation hearings.

Biden had tried unsuccessfully for the Democratic presidential nomination in 1988 and 2008, but won the nomination in 2020. He chose Kamala Harris to be his running mate, and Biden and Harris defeated incumbent president Donald Trump and vice president Mike Pence in the 2020 election. Upon election, Harris became the first woman and the first person of color to occupy the vice presidency.

The corona virus pandemic was raging during the 2020 election year. In his campaign, Biden promised to utilize the Defense Production Act and to make 100 million vaccines available during his first one hundred days in office. He fulfilled that promise.

Biden's early presidential activity focused on helping the United States recover from the COVID-19 pandemic and the recession it had caused. He proposed, supported, and signed into law the American Rescue Plan Act of 2021. Biden also issued a series of executive orders to reverse several Trump administration policies. Significant among these actions was rejoining the Paris Agreement on climate change. He completed the withdrawal of U.S. troops from Afghanistan by September 2021. During the withdrawal, the Afghan government collapsed and the Taliban seized control. There is divided opinion on the effectiveness of the withdrawal, but in some quarters, it has been called one of the most efficient military operations.

During and after the campaign, Biden had used the slogan "Build Back Better," and this characterized the bipartisan Infrastructure Investment and Jobs Act, which Biden signed into law in November 2021. In February 2022 Vladimir Putin's Russian forces invaded Ukraine. In response to this aggression, Biden led the international community is a series of sanctions against Russia, and authorized over $1 billion in U.S. weapons shipments to Ukraine. The efforts to thwart Putin's aggression are ongoing as of this writing.

Issues and Themes that Affect our Lives in Era 210

The Rise and Prevalence of Dystopian Literature

In our discussion of Era 203 (Seneca Falls Convention, Manifest Destiny, Gold Rush) we considered the effects of utopian literature. In the mid-1800s there was a sense among many people that human beings and human institutions were perfectible. There were at least a hundred communities across the country, planned by people who held those idealistic beliefs. In addition, many authors formulated ideas about ideal societies, and wrote novels about their versions of utopia. However, in the current era, utopian ideas and fiction have given way to a more pessimistic view of human society.

Dystopian literature is a genre of fictional writing used to explore social and political structures in a dark, nightmarish world. Whereas utopian literature explores the idea of societies that are positive, or even ideal, dystopian literature explores societies characterized by poverty, squalor or oppression. Dystopian themes are most used in science fiction and speculative fiction genres. When fears and anxieties about the future are prevalent, dystopian literature may serve the purpose of helping readers face and work through their fears.

Two of the most famous works of dystopian fiction were published in 1932 and 1949. They were Aldous Huxley's Brave New World and George Orwell's 1984. Both Huxley and Orwell explored the idea of societies existing with immense governmental control. I am including the discussion of dystopian literature in Era 210 instead of Era 6 (Roaring Twenties, Prohibition, The Great Depression, World War II) or Era 7 (Economic Recovery, World War II, The Truman Doctrine) because it is in Era 210 that dystopian literature has become the most popular form of fiction among young adult readers.

Professor Pamela Bedore of the University of Connecticut cites five areas of fear and anxiety experienced by young people that are explored in dystopian literature: liberty and choice, justice, environment, technology, and economic security. There are dozens of authors and scores of works that explore these themes. These themes are clearly at the heart of issues that society is facing. That dystopian literature is the most popular genre for young adult readers is an indicator of the anxieties that young people are experiencing as the future unfolds.

In 2008 Suzanne Collins published the first of a series: The Hunger Games. Collins' work will be briefly discussed since The Hunger Games became a television program and is therefore widely recognized. The story describes a world with a capital and twelve districts. Once per year twelve teams – one from each district – of a teenaged boy and girl, are randomly chosen and sent to participate in the hunger games. The games are televised and are required viewing for all citizens. The contestants are sent to a dangerous locale and eliminated by death until one remains. Death can be from random avalanches set off by the game-makers, or at the hands of other contestants.

That Collins was writing about war was very apparent in the third novel of her series, entitled Mockingjay. In that story there is a real war as the twelve districts are in rebellion. The story is complete with behavior modification of a captured soldier and memory modification methods. Collins explores issues of expedience, sexuality, and reluctance to procreate in a dystopian society.

The Hunger Games is but one example of the multitude of works in the genre of dystopian fiction. The spirit of our times is one in which fear of a very non-utopian future is prevalent.

<u>An Age of Bread and Circuses</u>

The term "bread and circuses" was coined to describe an era in the history of the Roman Empire. The empire had expanded in a period of growth and vitality that lasted a few centuries. During its vital, expansive phase Rome was governed by a democratic system that was not unlike ours in that there were chief executives called "Consuls," and there was a legislature that they called the Senate. In the year 60 BCE the democratic system was replaced by three powerful men, including Julius Caesar, who called their ruling system a "triumvirate." Thirty-three years later Augustus became the first of many Roman emperors. The Roman populace lost its vitality, though the empire would survive for five centuries. Four of those centuries were characterized in part by the famous (or infamous) gladiatorial games and chariot races. The rich and powerful rulers maintained their power by pacifying the masses. The Roman poet and satirist Decimus Juvenalis is quoted as having written, "The people that once bestowed commands, consulships, legions, and all else, now concerns itself no more, and longs eagerly for just two things – bread and circuses." Thus, "Bread and Circuses" refers to a diet of entertainment or political policies on which the masses are fed to keep them happy and docile.

Era 210 in America bears some disquieting resemblance to ancient Rome. Beginning soon after the Revolution, the United States entered an era of geographic expansion, military power, economic strength, technical inventions, public education, establishment of a huge college and university system, and citizen participation in government. Our history has been far from perfect, as you have read in the history contained in this book. Nonetheless it can be argued that there has been considerable vitality in the United States for two hundred years.

Have the past forty years been as vital? Let's look at a few trends. Until the end of the 1960s a fan of professional football could watch the sport on Sunday. Beginning in 1970, Monday night football became available. With today's various television networks there are now opportunities for a fan to see every single game. Until the 1970s television watchers were limited to a small number of local channels. In the 1970's cable television expanded into major cities, and with it came a multitude of channels. A Nielsen report indicates that United States adults are watching over thirty-five hours of television per week. This adds up to slightly more than 77 days per year. Home computers became common in the 1980s, and with them came video games. According to Nielsen, the average hours per week Americans play video games rose from 5.1 in 2011 to 5.6 in 2012 and to 6.3 in 2013.

Immersion in an electronic screen is being presented not only as acceptable, but as admirable. As these words were written, there was a current television advertisement that showed a man walking his dog and playing with his hand-held electronic device. The advertisement told us that this man was able to watch his favorite sports and programs all the time, even while walking his dog. The man was not watching to see if his dog were happy. He was not looking at nature. He was not looking at, nodding to or smiling at passers-by. He was not even watching for cars as he crossed streets.

These comments are but a tiny snippet of the myriad ways in which Americans are becoming immersed in their electronic devices and entertainment, and the ways in which this immersion is being legitimized and encouraged. With all this immersion in electronic entertainment, how much time is left for the following?

- Education

- Physical fitness

- Keeping up with political events

- Keeping up with economic trends

- Keeping up with social issues

If people are not keeping up with social, economic and political issues, how do they form the opinions by which they vote? They may not have time or inclination to educate themselves and thoughtfully construct their views. They are more susceptible to slogans, impressions, sound bites and public relations campaigns.

Another advertisement that I believe is significant showed a service being promoted. At a meeting of advertising staff, a tall, confident man recommended saying, "Baddabook baddaboom." When a more informative approach was suggested, he replied, "Nope! Just say Baddabook baddaboom!" What was the message of this advertisement? The message appears to have been: If a person who appears competent and confident says something, you may believe it, even if it is utter nonsense.

The public's immersion in electronic devices and entertainment, combined with the idea that you can believe nonsense if it comes from a confident source, is a recipe for extraordinarily ill-informed opinion formation and voting behavior. This is an extremely disquieting trend in the life of our democratic republic.

The Corona Virus Pandemic of 2020

In December of 2019, an individual in Wuhan, China became ill with a virus that we now know as Covid-19. Worldwide commerce and travel made it inevitable that the disease would spread. On January 20, 2020, the first case of corona virus in the U.S. was confirmed by a laboratory. The stricken individual had just returned to Washington State from Wuhan. Even prior to that confirmed case, the U.S. Center for Disease Control and Prevention (CDC) had established an "incident management structure." The CDC informed the administration of the seriousness of the threat and began preparations to help local jurisdictions cope with a likely epidemic.

As early as January 2020 President Trump was being warned in security briefings about the potential seriousness of the disease. On January 22, Trump made the following statement to the press: "We have it totally under control. It's one person coming in from China, and we have it under control. It's going to be just fine." On January 24, Trump issued the following statement: "China has been working very hard to contain the Coronavirus. The United States greatly appreciates their efforts and

328

transparency. It will all work out well. In particular, on behalf of the American People, I want to thank President Xi!"

On January 30, the coronavirus was declared a global health emergency by the World Health Organization. On that same sate, Trump stated, "We have very little problem in this country at this moment — five [cases]. And those people are all recuperating successfully. But we're working very closely with China and other countries, and we think it's going to have a very good ending for us." The United States represents five percent of the world's population. However, throughout the Covid-19 pandemic the U.S. has consistently had twenty percent of the world's Covid-19 cases and twenty percent of the world's deaths from the disease.

Trump took a few window dressing measures, such as creating a task force with vice-president Pence at its head. But the task force did little other than giving daily press briefings. On February 27, 2020 Trump stated, "It's going to disappear. One day it's like a miracle, it will disappear." And the next day, while talking about the Democratic Party, he referred to the corona virus as "one of their hoaxes." Trump's lack of active leadership, plus those statements, set the tone for what followed. Throughout early March Trump continued to broadcast unwarranted optimism about the trajectory of the virus. He also promoted hydroxychloroquine as a treatment, even after the Food and Drug Administration warned that it can cause heart problems.

On March 11, The WHO officially labeled the coronavirus a pandemic. On March 13 Trump acceded to the need for a response and declared a national emergency, freeing up federal funds. The virus continued to spread, and on March 19 California became the first state to implement a general stay-at-home order. At a March 19 press conference, Trump labeled the coronavirus the "Chinese Virus"; photos showed that he revised prepared remarks to add the xenophobic term. By the end of March, more than 30 states had issued shelter at home orders. In the opinion of most experts, the shutdown was a public health necessity. They did slow the US economy.

The months ahead saw Trump railing out against certain (Democratic) state governors, issuing random accusations against various persons and agencies, including the World Health Organization, and fanning flames of discontent while offering no real leadership. By the last week of April, the U.S. reported one million Covid-19 cases and 50,000 Covid-19-related deaths.

On May 8 Trump claimed that the U.S. is "the world leader" in dealing to the coronavirus, despite evidence indicating the U.S. was among the three countries with the worst record. Both these trends continued: an increase in Covid-19 cases and deaths, and Trump's attempting to minimize the issue. By mid-2022, in the U.S. the corona virus had accounted for eighty-four million cases and 1,003,000 deaths

Inventions that Changed our Lives in Era 210

The Hubble Space Telescope

Until Galileo Galilei turned his telescope toward the skies in 1610, all humankind knew about the Universe came through observations with the naked eye. Beginning with Galileo, we learned that Saturn had rings, that Jupiter had moons, and that the cloudy area across the middle of the night sky called the Milky Way is a collection of countless stars. Within a few years, our understanding of the natural world, and of our place within it, had been forever changed.

In the following centuries, human beings built larger and more powerful telescopes. They were placed far from city lights and as far above the haze of the atmosphere as possible, on mountains such as California's Mount Palomar, Kitt Peak near Tucson, Arizona, Mauna Kea in Hawaii and Cerra Tololo in northern Chile. But as high and as dry as we could place our telescopes, our atmosphere always interfered with seeing.

If a telescope could be placed in orbit, above the atmosphere, its power would be vastly increased. In April of 1990, such a telescope was launched. It was named after Edwin Hubble, the astronomer whom we discussed in Era 6 (Roaring Twenties, Prohibition, The Great Depression) and who discovered that "spiral nebulae" are in fact distant galaxies and that the Universe is expanding. Far above the atmosphere, Hubble has an unobstructed view of the universe. Scientists have used Hubble to observe distant stars and galaxies as well as the planets in our solar system.

The April 1990 launch of Hubble was the most significant advance in astronomy since Galileo's telescope. Our view of the universe and our place within it has been forever enhanced. Incidentally, adaptive optics, a technology by which atmospheric disturbance is screened out, has now substantially improved ground-based observations as well.

Promise of Fusion Power

This is a potential invention that has not yet changed our lives, but that could do so in the foreseeable future. In Era 7 (Economic Recovery, World War II, The Truman Doctrine) we discussed how fission power was discovered. There is enormous energy stored in the chemical bonds of heavy elements such as uranium and plutonium. Fission power is produced by the breakdown of heavier elements. But energy is also produced when lighter elements are fused together to form heavier elements. Almost all the energy we experience on Earth is or was produced by the latter process – nuclear fusion.

As you read these words, the Sun is fusing hydrogen into helium and producing energy. The energy is emitted in the form of gamma radiation, the shortest and most energetic wave length of electromagnetic radiation. Every second, the Sun fuses perhaps fifty-five million tons of hydrogen into perhaps fifty million tons of helium. Five million tons of matter are converted into energy. The energy created is immense, as described by Einstein's famous equation: energy = mass x speed-of-light-squared, which we discussed in Era 5 (Telegraph, Populist Movement, Pure Food and Drug Act).

Nuclear fusion - the production of energy from the fusion of hydrogen nuclei - has been termed "the holy grail of energy." It would be incredibly productive and largely clean. But a method to contain the plasma (protons without attendant electrons) has not been devised, and harnessing nuclear fusion seems always to be "a few decades away." However, in October of 2014, the Lockheed Martin corporation announced it had achieved a technological breakthrough, and they claimed they will be able to make a compact fusion reactor within a decade. The plasma will be controlled by a process they call "magnetic mirror confinement." Without details on the process, some outside scientists are skeptical. But if Lockheed were to produce workable nuclear fusion, the world of energy production would never be the same.

The Discovery and Promise of Graphene

Graphite is a crystalline form of the element carbon with its atoms arranged in a hexagonal structure. The properties of graphite have been explored for more than a century and a half. In 1859 Benjamin Collins Brodie recognized that graphite oxide is arranged in fine sheets of material, known as a lamellar structure. In 1947 Philip Russell Wallace conceived the theory of graphene.

It was a half a century before Graphene was first created by Andre Geim and Kostya Novoselov at the University of Manchester. They were able to peel off layer after layer of graphite, until they had a single layer of carbon atoms in a two-dimensional honeycomb structure. They analyzed their creation and found the graphene that Wallace had theorized. They published their discovery in 2004 and won the 2010 Nobel Prize in Physics.

Graphene is a material that holds enormous promise for industrial and commercial use. In its perfect crystalline form, it has been called the strongest material ever measured. Graphene has an extraordinary combination of properties. It is optically transparent and is a superb conductor of heat and electricity. Though it has been in use for only a few years, it is already making its presence felt in the areas of batteries, smartphones, energy storage and sports equipment. It is said to have potential applications in solar panels, medicine, sensors, electronics, flexible display screens, and anti-corrosion paints and coatings. We may well have caught only the initial glimpse of the transformative power of this super-thin, strong, conductive and all-round surprising substance.

Medical Advances in Era 210

<u>Mapping of the Human Genome</u>

In our every cell there is a blueprint for the structure and function of every aspect of our bodies. This information is contained in twenty-three sets of chromosomes. Each chromosome contains over a thousand genes. Genes are composed of nucleic acids, which are one of the four basic molecules of which life is composed: sugars, nucleic acids, amino acids and hydrocarbons (fats). Nucleic acids are composed of a five-carbon sugar (usually ribose nucleic acid or deoxyribose nucleic acid), a phosphate group (a phosphorus atom surrounded by four oxygen atoms) and a base. The four bases are adenine (A), guanine (G), cytosine (C) and thymine (T). This information is but a scratch on the surface of a sphere of information that your author cannot begin to fathom. Suffice it to say that our genetic make-up is extremely complex. But within that make-up is information with enormous potential to advance our understanding of and treatment of human medical problems.

In 2000, scientists with the Government's Human Genome Project and those from a private company, Celera Genomics, released a rough draft of the human genome to the public. In 2003 a "final draft" was released by researchers, and in 2007 more updates to the genome were published. For the first time the world could read the complete set of human genetic information. This would set the stage for discovery of the specific functions of our roughly 23,000 coding genes.

The mapping of the human genome essentially meant that all the "words" were identified. The "sentences, paragraphs and meaning" were in reach, but were yet to be deciphered. Those advances are beginning to come, and scientists predict it will help both preventative and restorative medicine.

Many medicines are effective for only a small percentage of patients. This is due in great part to genetic variations among people, variations that determine how a medicine is metabolized and cleared by the patient. Even now, there are companies that have developed methods to help select the best treatments for maladies, such as different types of cancer, by matching the treatment with the patient's genetic make-up. This expanding body of research and knowledge, known as "precision medicine," has huge potential to help relieve suffering and to cut our rapidly accelerating medical costs from increasingly expensive treatments – treatments that often are a mismatch with the patient's genetics.

Antibiotics and Drug Resistant Pathogens

This discussion is of a medical concern as opposed to a medical advance. It is an issue that has been gaining increased attention in recent years.

Antibiotics have been in common use for the past 70 years to treat infectious diseases. Diseases that would previously have caused prolonged discomfort or death have been rendered much less of a problem since the 1950s. However, the frequent use of these medications over so long a time period has influenced the virulence of the organisms the antibiotics are designed to kill. This is due to the adaptability of the bacteria and other pathogens. If we manage to kill ninety percent of the bacteria in any sample, the ten percent that remain are those that are less susceptible to the agent that killed their relatives. If the survivors reproduce to its former population level, this population will have been produced by the resistant strains. Repeated cycles of this type have created what some call "superbugs" – pathogens that are resistant to many if not all the agents physicians have available to combat them.

Each year at least two million people in the United States become infected with bacteria that are resistant to antibiotics. At least 23,000 people each year die as a direct result of these infections.

We have all heard of "staph infections," which are a common problem in hospitals. The pathogen is called staphylococcus aureus. When penicillin was introduced, it was effective against staphylococcus aureus. Within ten years staph had become resistant to penicillin, and an antibiotic called methicillin was introduced. Within several years staph had become resistant, and the resulting pathogen is called "methicillin resistant staphylococcus aureus," or MRSA. Vancomycin was then introduced, and it worked until staph was able to adapt a strain called VRSA, or "vancomycin resistant staphylococcus aureus." This is a well-known example, but only one example, of an "arms race" that medicine is fighting against a well-equipped adversary. With what is the adversary equipped? Our pathogenic adversaries are equipped with the ability to reproduce with speed. It takes nine months for a human couple to make a newborn, while it takes a bacterium such as *e coli* twenty minutes to create a new generation. That kind of rapid reproduction gives bacteria a huge advantage in the arms race.

What We Learned About the Universe in Era 210

The Kuiper Belt

Gerard Kuiper was a Dutch-American astronomer who helped transform our knowledge of our Solar System. Kuiper received his Ph.D. at the University of Leiden in 1933, moved to the United States, and joined the staff of Yerkes Observatory of the

University of Chicago in 1936. Kuiper founded the Lunar and Planetary Laboratory at the University of Arizona in 1960 and served as its director until his death.

Kuiper has an impressive research record. In 1944 he confirmed the presence of methane in the atmosphere of Saturn's moon Titan. His 1947 prediction that carbon dioxide is a major component of the atmosphere of Mars turned out to be correct, as did his prediction that the rings of Saturn are composed of particles of ice. He discovered the fifth moon of Uranus and the second moon of Neptune. In 1950 he obtained the first reliable measurement of the diameter of Pluto. In 1956 he proved that Mars's polar icecaps are composed of frozen water, not of carbon dioxide.

In 1949 Kuiper formulated a theory of the origin of the solar system. He suggested that large cloud of gas contracted to form the Sun, and that the planets had formed by the condensation of the gas and dust left in orbit about the young Sun. He also suggested the possible existence of a disk-shaped belt of comets orbiting the Sun at a distance of 30 to 50 astronomical units (one astronomical unit is the distance from the Sun to the Earth – 93 million miles). The existence of this belt of millions of comets was verified in the 1990s, and it was named the Kuiper Belt. The Solar System had become more complex and interesting.

Discovery of Kuiper Belt Objects

Michael Brown is well known in the scientific community for his surveys for distant objects orbiting the Sun. His team has discovered many trans-Neptunian objects (TNOs). Particularly notable are Eris, Haumea, Makemake, Quaoar, Sedna, and Orcus. Eris is classified as a dwarf planet and the only TNO known to be more massive than Pluto. Sedna is a planetoid thought to be the first observed body of the inner Oort cloud, a cloud of comets from fifty to a thousand astronomical units from the Sun. The Solar System appears more complex and varied to us as astronomers continue to make discoveries.

Discovery of Extrasolar Planets

Astronomers have believed for a long time that there are planets in orbit about other stars. To believe otherwise would be to assert that our solar system is unique. Based on the theory of how our solar system was formed, debris surrounding stars in the process of formation is expected to coalesce (the term used is "accrete") into larger and larger objects until some planetary bodies are present.

The tools to identify extrasolar planets have only recently become available. The first such object was discovered in 1995 by Swiss astronomers Didier Queloz and Michel Mayor. They discovered a planet in the constellation Pegasus and dubbed it 51 Pegasi b.

Extrasolar planets cannot be seen visually as they are lost in the intense light of the stars that they orbit. They can be detected in a couple of ways. If the planet crosses in front of its star from our line of sight, there is an extremely small, but regular and detectible, reduction of the brightness of the star. Alternatively, a massive (Jupiter-sized) planet would exert a small gravitational tug on its star. This can be detected by an extremely slight, alternating redshift and blueshift in the star's light.

More than 4,850 extrasolar planets are now known to exist. As astronomers' methods become increasingly more sensitive, smaller planets will be detected. Planets only two-to-three times the mass of the Earth have now been discovered.

Advances in Origins-of-Life Research

As indicated in the discussion of Era 207 (The Transistor, Penicillin), in 1969 a large meteorite weighing more than forty pounds fell to earth near Murchison, Victoria, in Australia. It is a member of a class of meteorites known as carbonaceous chondrites - stony bodies with carbon compounds. It became a highly studied meteorite because it contained 3.5 percent by weight of organic compounds. David Diemer of the University of California received a piece of the meteorite for study.

As is widely known, there is a tendency in the Universe called "entropy." This describes the tendency in the Universe for energy systems to go from order to disorder. However, the fact of our existence argues that there is another tendency in the Universe for energy systems to become more organized. Scientists have recently been referring to this tendency as "emergence."

Through a series of careful steps, David Diemer extracted lipid molecules from the meteorite. Diemer placed the lipids in water and saw emergence in action as lipid molecules that had been formed in the vacuum of space billions of years ago organized themselves into vesicles with a bi-layered structure, a similar structure to that employed by the cell membranes in our bodies and in all life on Earth.

Diemer stated, "The self-assembly process seems to defy our intuitive expectation from the laws of physics that everything, on average, becomes more disordered." Diemer published his findings in 1989.

The search for life's origins is far from over. Vesicles made of self-organizing lipid molecules are suggestive of an origin of cell membranes. But remember that cell membranes are complex structures that must transport materials in and out of the cell using myriad protein receptors. The full story of the mystery of life's origin is still not solved. But if a vesicle made of self-organizing lipids were to capture inside of it a suite of the organic molecules that were undoubtedly available in quantity on the primordial Earth, then we could imagine a series of steps that could lead to life.

Space Exploration in Era 210

<u>Roving the Surface of Mars</u>

Over the years, NASA has sent five robotic vehicles, called rovers, to Mars. The names of the five rovers are: Sojourner, Spirit, Opportunity, Curiosity, and Perseverance

On January 4 and 25, 2004, two exploratory vehicles called *Spirit* and *Opportunity* landed on opposite sides of Mars. These two rovers were more mobile than their predecessor, *Sojourner*. These robotic explorers are operated by mission control on Earth, and are maneuvered very slowly to avoid disabling accidents. They have nonetheless trekked for miles across the Martian surface, and have provided information about Mars' atmosphere and geology. The rovers have found evidence that wet and habitable conditions existed intermittently in ancient Martian environments.

On 30 July 2020 a new Mars rover named *Perseverance* was launched. Perseverance is an upgraded version of its predecessor, *Curiosity*. It is the most sophisticated planetary research vehicle to date, and it landed successfully on Mars in February 2021, approximately 210 days after launch. *Perseverance* is the size of a small car and carries nineteen cameras and seven scientific instruments. Nine of its nineteen cameras are devoted to safe navigation. The rover also carried an extraordinary innovation: a mini-helicopter called *Ingenuity*. Mars' atmosphere is only $1/100^{th}$ the density of Earth's, so attaining lift and flight is a challenge. Nonetheless *Ingenuity* made the first powered flight on another planet on 19 April 2021, and has made 19 flights as of this writing.

Astronomers and planetary scientists have long been dedicated to searching for signs of life other than on Earth. *Perseverance* was therefore designed to explore the crater Jezero, as evidence indicates that area was once flooded with water and had an ancient river delta. *Perseverance's* mission includes seeking evidence of former microbial life, collecting rock and soil samples to store on the Martian surface for later collection, and finding out if oxygen can be extracted from the Mars' thin atmosphere. As of 21 April 2022, Perseverance has been active on Mars for 427 Earth days.

The Galileo Mission to Jupiter

On October 18, 1989 NASA launched the Galileo mission from the Kennedy Space Center in Florida. Galileo received gravity boosts, first from Venus and then from Earth, on its way to the Solar System's largest planet, Jupiter. We already knew, as a result of NASA's 1962 Mariner 2 flyby, that Venus has a surface temperature of roughly 900 degrees. A dozen and a half subsequent Venus missions had enabled scientists to map Venus' surface and explore its atmosphere, of carbon dioxide, with clouds of sulfuric acid. After its gravity assists from Venus and Earth, Galileo became the first spacecraft to visit an asteroid, when it flew by two such objects: Gaspra and Ida. The principle aim of Galileo was to study Jupiter and its mysterious moons, which it did in spectacular fashion. Galileo found that the Solar System's largest moon, Ganymede, generates a magnetic field. We also learned that Jupiter's moon Io is the Solar System's most volcanic body. Its volcanoes repeatedly and rapidly resurface the moon. Europa, another Jovian moon, is one of the Solar System's most intriguing bodies. It is covered with ice that is strongly believed to have a liquid or slushy saltwater ocean beneath it. Scientists are curious as to whether, at the bottom of that ocean, there could be hydrothermal vents of the type found on our own ocean floors. If so, could that environment harbor life?

Galileo orbited Jupiter for almost eight years, making close passes by all its major moons. It even sent a small probe deep into the Jovian atmosphere. The probe was crushed by overwhelming pressure, but not before taking readings for almost an hour. Galileo had insufficient power to escape the immense gravity of the Jovian system. To avoid a possible crash into Europa, potentially contaminating any life there, Galileo was allowed to plunge into Jupiter's atmosphere in September 2003 after a scientifically remarkable mission.

The New Horizons Mission to Pluto and Beyond

After its discovery by Clyde Tombaugh in 1930, Pluto was seen as the Solar System's ninth planet. In 2006 the International Astronomical Union revised Pluto's status to "dwarf planet," a designation it shares with the asteroid Ceres and with Kuiper Belt objects Eris, Haumea and Makemake. Though probably best seen as a Kuiper Belt object, Pluto's designation is not as important as the role it and other Kuiper Belt objects play in our developing understanding of the Solar System's origins and evolution. In fact, exploration of the Kuiper Belt – including Pluto – is highly ranked in importance by the National Academy of Sciences.

In January 2006 the New Horizons spacecraft was launched. It would not reach Pluto for nine years. The craft swung by Jupiter in February 2007 for a gravity boost, and flew by Pluto in the summer of 2015. For six months it took pictures and made measurements of Pluto and its companions. It had been known that Pluto has a large (compared to its own size) moon, Charon. But four new moons were detected: Kerberos, Hydra, Nix and Styx. Missions in the Solar System always find more complexity and wonder than anticipated. Pluto is not a nondescript body, but is a complex world of mountains, ridges and valleys, mostly sculpted of methane ice. Having flown by Pluto, the New Horizons spacecraft will journey farther into the Kuiper Belt to examine another ancient, small, icy world in that vast region.

The Restless Earth in Era 210

The 2013 Chelyabinsk Meteor

On February 15, 2013, an extremely bright light flashed across the sky above Chelyabinsk, Russia. It was a large meteor, and because it came in on a shallow angle, it was seen traveling a long distance across the sky. When large meteors enter Earth's atmosphere, they do so at extremely high speeds and air pressure builds up enormously in front of them. This causes them to "pancake," and often explode above Earth's surface. This was the case with the Chelyabinsk meteor. It is estimated it was traveling 12 miles per second and it exploded about eighteen miles above ground, generating a bright flash and a shock wave.

The shock wave caused windows to break, and damaged an estimated 7,200 buildings in six cities across the region. Authorities had to work quickly to help repair the structures in sub-freezing temperatures. About 1,500 people were injured seriously enough to seek medical treatment. Most of the injuries were due to broken glass.

It is fortunate that most of the object's energy was absorbed by the atmosphere. An object of that mass and traveling at that speed would carry the kinetic energy, before atmospheric impact, of 400–500 kilotons of TNT - between 26 and 33 times as much energy as the atomic bomb detonated at Hiroshima.

The Chelyabinsk meteor is estimated to have had an initial mass of about 12,000–13,000 metric tons (heavier than the Eiffel Tower), and to have been about 20 meters in diameter. It is the largest known natural object to enter Earth's atmosphere since the 1908 Tunguska event described in Era 5 (Telegraph, Populist Movement, Pure Food and Drug Act) that destroyed a wide, remote, forested, and sparsely populated area of Siberia. The Chelyabinsk meteor is also the only meteor confirmed to have resulted in many injuries. No deaths were reported.

The Chelyabinsk incident is a reminder that there are many thousands of objects in near-Earth orbit whose orbits could potentially intersect Earth's orbit at some point in the future.

<u>The Eruption of Mount Pinatubo</u>

The second-largest volcanic eruption of this century, and by far the largest eruption to affect a densely populated area, occurred at Mount Pinatubo in the Philippines on June 15, 1991. The eruption produced pyroclastic flows, which are high-speed avalanches of hot ash and gas. It also produced lahars, which are giant mudflows, as well as a cloud of volcanic ash hundreds of miles across. The impacts of the eruption continue to this day.

Mount Pinatubo had been inactive for five hundred years, but the June, 1991 eruption was not without considerable warning signs. On July 16, 1990, a magnitude 7.8 earthquake struck about 60 miles northeast of Mount Pinatubo on the island of Luzon. That quake was similar in size to the great 1906 San Francisco, California, earthquake. At Mount Pinatubo, this major earthquake caused a landslide, some local shaking, and an increase in steam emissions. In March and April 1991 molten rock rising toward the surface triggered small earthquakes and caused powerful steam explosions on the north flank of the volcano. Thousands of small earthquakes occurred beneath Pinatubo through April, May, and early June, and many thousand tons of noxious sulfur dioxide gas were emitted by the volcano.

From June 7 to 12, magma reached the surface of Mount Pinatubo. However, it had lost most of its gas on the way to the surface, so the magma oozed out without an explosive eruption. But highly gas-charged magma reached Pinatubo's surface on June 15, and the volcano exploded in a cataclysmic eruption that ejected more than 1 cubic mile of material. The ash cloud from the giant eruption rose 22 miles into the air. A blanket of volcanic ash covered the countryside. Fine ash fell as far away as the Indian Ocean, and satellites tracked the ash cloud several times around the globe. So much magma and rock were blasted out from below the volcano that the summit collapsed to form a caldera 1.6 mile across.

Climate science had evolved to the point that scientists were able to use the eruption as a natural experiment. Predictions were made of the likely effect of the eruption on worldwide temperatures. A one-degree drop was predicted, and the prediction was verified.

Hurricane Katrina

Hurricane Katrina originated as a tropical depression over the Bahamas on August 23, 2005. The following day, it intensified into a tropical storm. The storm strengthened into a hurricane two hours before making landfall on Florida's east coast. It weakened somewhat as it crossed Florida, but quickly strengthened to hurricane force as it emerged into the Gulf of Mexico. The storm strengthened to a Category 5 hurricane over the warm waters of the Gulf of Mexico, and made its second landfall as a Category 3 hurricane on August 29, in southeast Louisiana.

The storm caused catastrophic damage along the Gulf coast from central Florida to Texas, much of it due to the storm surge and levee failure. Severe property damage occurred in coastal areas. In Mississippi, water reached 6–12 miles from the beach. Overall, at least eighteen hundred people died due the hurricane and to subsequent floods and disease, making it the deadliest United States hurricane since the 1928 Okeechobee hurricane.

New Orleans' hurricane surge protection failed. There were more than fifty breaches in the system, many on the levee system to the north of the city. Eighty percent of the city was flooded, and floodwaters lingered for weeks. According to the U.S. Army Corps of Engineers, two-thirds of the deaths in Greater New Orleans were due to levee and floodwall failure, as well as to reported FEMA ineptitude.

Hurricane Maria

The worst natural disaster to ever hit the islands of Dominica, the Virgin Islands and Puerto Rico was Hurricane Maria, which struck in September 2017. Maria was a Category Five storm. It brought massive rains that caused flooding, catastrophic destruction and numerous fatalities. The infrastructure of the island was already considered to be shaky after years of neglect, and roads, bridges, electrical power and water supply were all massively impacted.

The death toll from the storm continued to rise for months, as deaths due to disease continued to be reported. Ten months after the storm, the island's governor reported a death toll of 2,975.

Hurricane Maria is remembered due to politics as well as due to meteorology. There were many reports of ineffective, inefficient and apparently unmotivated disaster response and humanitarian aid. President Trump stated that the federal government did "a fantastic job" responding to the storm, despite considerable evidence to the contrary. Trump repeatedly stated, "I give us a ten."

<u>Hurricane Harvey</u>

Hurricane Harvey made landfall three times in Texas and Louisiana between August 25 and August 30 of 2018. On August 25 Harvey made landfall as a category 4 hurricane with 130 mile/hour winds and did cause a massive storm surge that flooded coastal areas. But the major damage was caused after Harvey's winds had died down to around forty miles/hour. The storm stalled and dumped a typical year's rain in less than a week on much of southern Texas, Louisiana and southern Arkansas. More than fifty inches of rain fell on parts of Houston. Harvey was the second most costly hurricane to hit the United States since 1900, trailing only Katrina. Thirteen million people were affected and almost 135,000 homes were damaged or destroyed.

Miscellaneous Events in Era 210

<u>Has the Foreign Policy of Containment been Effective?</u>

As has been described in earlier sections, the policy of containment of the spread of communism has been the defining feature of United States foreign policy since it was devised by President Harry Truman and his Secretary of State John Marshall after World War II. Has it been effective? As of 2013, forty-five percent of the world's almost two hundred politically separate entities and forty percent of the world's population are living in freedom. No attempt is made here to estimate what the percentage of the world's population would have been living in freedom without the policy of containment. A perusal of articles on the subject reveals differing opinions and examples of presumed successes and failures. Will human civilization benefit from the U.S. involving itself in other parts of the world when anti-democratic factions assume power? These are questions worthy of consideration, since the policy has cost the nation hundreds of thousands of lives and trillions of dollars.

Intergovernmental Panel on Climate Change

Thousands of scientists around the world have been studying the Earth's climate system for decades. Earth's climate system is complex and involves all the Earth's major systems: the atmosphere, biosphere, hydrosphere (water), lithosphere (rocks), and cryosphere (ice). The study of how all these systems interact to affect one another is sometimes referred to as "Earth systems science." In order to provide as much efficiency as possible in reporting findings to governmental policymakers, the Intergovernmental Panel on Climate Change (IPCC) was established in 1988 by the World Meteorological Organization (WMO) and the United Nations Environment Program (UNEP) to assess and report on climate change based on the latest science.

Every five to seven years, through the IPCC, thousands of experts from around the world synthesize the most recent developments in climate science, and governments can request access to the reports. The content of IPCC reports is policy-relevant, but not policy-prescriptive. Government representatives work with experts to produce the "summary for policymakers" (SPM) that highlights critical developments in language accessible to the world's political leaders. The original scientific reports are available to scholars, academics and students seeking a deeper understanding of the evidence. The IPCC has issued comprehensive assessments in 1990, 1996, 2001, 2007 and 2013.

The 1990 report indicated that if carbon dioxide levels in the atmosphere were to rise to twice pre-industrial levels, there would be a dangerous anthropogenic (i.e. human-induced) interference with the climate system. The 2007 Assessment Report emphasized that "the warming of the climate system is unequivocal" and that it is affecting ecosystems worldwide. And the 2003 Assessment Report stated that "human influence on the climate system is clear, and recent anthropogenic emissions of greenhouse gases are the highest in history." These findings led to the Paris Agreement of 2015, in which 197 countries committed to limiting global warming. The United States was a participant in the Paris Agreement until Donald Trump withdrew us from that agreement in June of 2017. President Joe Biden has reinstated our participation.

The IPCC produces reports that support the United Nations Framework Convention on Climate Change (UNFCCC), an international treaty on climate change. The ultimate objective of the UNFCCC is to "stabilize greenhouse gas concentrations in the atmosphere at a level that would prevent dangerous anthropogenic interference with the climate system."

Summary Comments on Era 210

President Bill Clinton took office after George H. W. Bush. The domestic programs enacted under Clinton are significant. The North American Free Trade Agreement signed by President Bush was enacted. Clinton is the last President to have balanced the federal budget, having done so for three consecutive years, 1998 through 2000, and he presided over the longest period of peacetime economic expansion in American history. Tax cuts were enacted for 98.8 percent of taxpayers. Only those who have benefited the very most from our economy – the top 1.2 percent of earners – were required to pay higher taxes.

Clinton championed the Family Medical Leave Act and the Brady Handgun Violence Prevention Act. By the end of his second term crime was at a twenty-five-year low and the unemployment rate was at a thirty-year low. Clinton showed that government can respond to the needs of the people while still maintaining fiscal discipline.

A sexual indiscretion led to Clinton's impeachment and a near miss from conviction. People believed he sincerely tried to improve the nation and all its citizens rather than serving special interests or a small portion of the populace. No modern President has left office with a higher approval rating. Gallup polls showed Clinton was regarded by thirteen percent of Americans as the greatest President in U.S. history.

George W. Bush lost the popular vote to Al Gore in 2000 but won a narrow and disputed victory in the electoral college. From his predecessor, Bush inherited a healthy economy in which there had been federal budget surpluses three consecutive years. Bush did not maintain those trends. He returned to Reaganomics and granted a huge tax break to the wealthiest Americans. That action plus a disastrous war in Iraq sent the nation's budget deficit on a steep upward trend from which it has not recovered.

The nation was attacked on September 11, 2001. Bush responded with the Patriot Act, expanding government's ability to find and arrest people. Though aimed at terrorists, the act increases the government's ability to invade the privacy of all citizens.

Bush turned his attention to overthrowing Iraq's dictator, Saddam Hussein. Whether he wanted revenge on Hussein for an attempted assassination of his father, or whether he wanted to complete his father's unfinished work by removing Hussein is uncertain. What is certain is that there was little multinational support for an invasion. Nevertheless, with minimal support by other nations he invaded Iraq. Hussein was captured and was subsequently executed by the Iraqis.

Though successful in leading to Saddam Hussein's demise, the invasion of Iraq led to frustrating foreign policy in Iraq and difficult political and economic life at home. Many American soldiers lost their lives and the military expenditures were immense. The economy suffered, there were scandals within the Republican Party and Bush's approval ratings plummeted.

The American and world economies suffered after the 2008–2009 real estate credit crisis here in the U.S. Whether Bush's administration could have discerned the greed, collusion and excesses that caused the crisis is debatable. However, the signs were there that lending institutions were providing extremely risky loans, so it is probable that Bush's administration could have taken steps to avert the crisis. George W. Bush left the nation with an escalating national debt, and more divided and insecure than he had found it.

Barak Obama succeeded George W. Bush and inherited a difficult economic situation. After the bank debacle of 2007-2008, former President Bush had provided a huge, controversial bailout for the banks. Obama's team worked closely with Bush's outgoing administration and it did not appear prudent for Obama to try to reverse the course on which Bush had set us. Thus, an economic stimulus package was Obama's first order of business, and the quantitative easing program was put in motion. The economy did grow at a steady though not fast rate throughout Obama's administration.

The American tradition of health insurance being tied to employment was an historical accident. The U.S.'s dubious distinction of being the world's only major industrialized nation without universal healthcare was finally corrected in 2010 by the passage of the Patient Protection and Affordable Care Act.

By August of 2010 Obama was close to his goal of withdrawing all U.S. troops from Iraq, but ISIS activity forced the U.S. to remain to assist Iraq's security forces. The plan had been to withdraw troops after eighteen months and to turn full responsibility for that country's security to the Afghans. Obama's goal of no American troops fighting in Iraq was reached in 2012.

In 2011 Obama authorized Navy SEALs to conduct a "surgical raid" on a compound in Pakistan where Osama bin Laden was in hiding. Bin Laden died in the raid, his identity was confirmed by DNA testing, and he was immediately buried at sea. Both parties at home and many overseas were glad that bin Laden had been dispatched.

Obama boosted the overseas reputation of the United States, and European polls indicated he was the most respected world leader. At home, he reminded us of what can be accomplished with an honest, well-reasoned approach to issues. His presidency was a force for human rights and environmental protection. Obama was ranked the eighth greatest American President in a 2018 poll of historians by the American Political Science Association.

President Obama was succeeded by Donald Trump. Trump lost the popular vote by approximately three million votes to Hillary Clinton, but nonetheless won the electoral college. It has come to light that there was considerable interference in our electoral process by Russian interests, much of which was ordered by Russian President Vladimir Putin. Much of the interference was conducted via the Internet and had targeted the states that were crucial to the election outcome.

Trump's presidency has been characterized by several trends. First, Trump has consistently endeavored to undo everything President Obama had attained. He withdrew us from the Paris Agreement on climate change and from the treaty to restrict nuclear arms development in Iraq. Trump tried as hard as he could to dismantle the Affordable Care Act (often known as "Obama Care"), though he fell a vote short in the Senate in his attempt.

Secondly, Trump supported and achieved a huge tax cut for corporations and for the very wealthiest Americans. This has caused another acceleration in the national debt, a trend begun by Ronald Reagan and re-ignited by George W, Bush. The supposed idea was that the corporate tax cut would encourage corporations to create more jobs and to increase wages. However, as many had predicted, corporations used the money primarily to buy back their own shares, thus raising their share prices. This benefits wealthy shareholders and the stock market, but does not achieve the goals of more jobs and better wages.

The most profound effect of the Trump presidency is the national mood and morale. The effect of Trump's rhetoric, during the election and throughout his tenure in office, has been to divide people. America's citizens and political parties are more sharply divided than ever. Trump's continual attacks on the press have served to reduce the public's trust in the media. This results in people being more inclined to form opinions based on preconceived notions, and on questionable social media and Internet sites, and to disregard information from traditional news media. There has been a sharp increase in hate-inspired crimes. Trump's rhetoric is consistently insulting and vitriolic, and many individuals and organizations have followed suit. This has resulted in a deterioration of the quality of public discussion of issues. Reasonable, honest and civil national discussion of issues has been seriously eroded during Trump's presidency. After losing the 2020 presidential election to Joseph Biden, Trump refused to concede the election. He attempted via sixty-two lawsuits to overturn the results in five states. Having lost his lawsuits due to lack of evidence, Trump invited white supremacist organizations to come to Washington, DC on January 6 to try to stop Congress from certifying the election results. The result was an insurrection in which the Capitol was violently invaded, Congressmen were endangered, and in which there were a few deaths. The event is widely seen as one of the darkest days in the history of the United States.

Joseph Biden succeeded Donald Trump as President. He fulfilled his promise of making one hundred million corona virus vaccines available in his first one hundred days in office. He appealed to both parties to help to "build back better" and in November 2021 he signed into law the bipartisan Infrastructure Investment and

Jobs Act. In February 2022 Russian President Vladimir Putin sent his military into Ukraine to subdue and annex that country. Biden led the international community in a series of sanctions against Russia, and authorized over $1 billion in U.S. weapons shipments to Ukraine. The efforts to thwart Putin's aggression are ongoing as of this writing.

Era 210 has seen advances in astronomy, such as the discovery of the Kuiper Belt, a realm of asteroids and dwarf planets beyond Neptune's orbit. Our knowledge about Mercury, Venus, Mars, and of Jupiter, Saturn, Uranus and Neptune and their many satellites has been revolutionized. The launching and operation of the Hubble Space Telescope has captured the imagination of many with its sharp images of celestial objects. We now know that planets do orbit other stars beyond our own.

Back on Earth, the Intergovernmental Panel on Climate Change, which integrates the research of thousands of climate scientists, issued a warning about dangerous anthropogenic interference with the climate system. Carbon dioxide emissions are a major issue, and carbon dioxide levels in the atmosphere are close to double pre-industrial levels. Ice sheets and glaciers are disappearing. The ranges of many species are changing. The frequency and severity of severe storms, floods and fires are increasing markedly.

If fusion power could be harnessed, it could be a relatively pollution free alternative to burning fossil fuels. In 2014 scientists at Lockheed Martin announced that within a decade they may be able to manufacture a small fusion reactor.

Another issue closer to home is our use of antibiotics to treat diseases. Antibiotics have been an enormous boon in alleviating human suffering and death. But the pathogens are "fighting back" with their ability to mutate, and treatment resistant strains (sometimes termed "superbugs") are evolving. Medical science faces a major challenge to determine the best path for us to take in our healthcare efforts. One exciting avenue for improvement in medical care is *precision medicine*, which is the tailoring of treatments to a patient's genetic structure. This advance is being made possible in the wake of the human genome project.

Era 210 saw the restless Earth in action. In 1991 Mount Pinatubo experienced an enormous volcanic eruption, and climate scientists made predictions that proved accurate about the effect on worldwide temperatures. In 2005 Hurricane Katrina crossed Florida and the Gulf of Mexico and slammed into Louisiana. Flood control barriers failed, the city of New Orleans flooded, and at least eighteen hundred persons perished. In 2013 a meteor the size of the Eiffel Tower entered the Earth's atmosphere on a shallow angle at twelve miles per second. It exploded over Chelyabinsk, Russia, damaging 7,200 buildings and causing injuries to fifteen hundred people. The Chelyabinsk meteor is a reminder that there are many objects in orbit about the Sun whose orbits may cross Earth's.

In 2017 Hurricane Maria devastated the American territory of Puerto Rico. In 2018 Hurricane Harvey stalled over Texas, dropping an immense amount of rain and causing unprecedented inland flooding.

Perhaps the most profound aspect of Era 210 is the effect that events have had on the human spirit. As utopian literature in the nineteenth century reflected society's optimism about human perfectibility and the future, twentieth and twenty-first century dystopian literature reflects society's fears and anxieties. Dystopian literature is the most popular genre among young adult readers, and the most common theme is fear of government curtailment of freedom. Optimism is not achieved by what appears to be a "bread and circuses" trend in our culture. Being able to see your favorite programs and sports at all times is portrayed as clever and desirable. This appears to be in direct conflict with encouraging people to stay aware of social, political and economic trends and to be informed voters and participants in societal decision-making.

Brief Summary of the Journey

In the introduction I used the analogy of this book as a two-and-a-half-mile walk, divided into ten segments, studying species of trees in the forest of American history. The species we would examine would be Presidents, inventions, medical advances, knowledge of the Universe, natural events and some miscellaneous species. Each quarter-mile we would stop and consider what we had seen. Having completed ten eras in our entire two-and-a-half-mile trek, what are some of the major trends and changes we have seen?

We began with a nation in which the only transportation was by horse. Canals and railroads were the first innovations; roads and automobiles were yet to come. We had had twenty-nine of our forty-five Presidents before automobiles were in common use in the 1920s. We now have an extensive interstate highway system, and we can travel by jet plane to any location in the world within a day.

Our nation began with communication no faster than the horse. Recall that the Battle of New Orleans was fought two weeks after the War of 1812 had ended, as news had reached neither the American nor British forces. The telegraph was the first major invention that sped up communication. Next came the telephone, and by 1900 there were more than a half-million telephones in use. In 1915, transcontinental calls were made possible. In our current era, most Americans walk around with a hand-held electronic device that can connect them with almost any of a few billion other phone users throughout the world.

During the nation's history there has been an array of inventions that have changed our lives. The harvesting of cotton was made much easier with Eli Whitney's invention of the cotton gin in 1793, and in that same year the Slater Mill, the country's first cotton spinning mill, was completed. Agriculture, the primary industry in the early years of the nation, was revolutionized by the McCormick Reaper, which came into general use in 1845. The unprecedented growth in the world's economy has been made possible by energy, and the first oil well was drilled in 1859. Henry Ford introduced the Model T in 1908, and the first assembly line in 1913. After exhaustive experimentation, Thomas Edison invented the electric lightbulb in 1879, and industry was no longer limited by the availability of sunlight. Reginald Fessenden made the first radio broadcast in 1900. In 1947 the transistor, the key active component in almost all modern electronics, was developed at Bell Laboratories. In 1952 IBM developed the first mainframe computer. Personal computers would be introduced by Apple in 1976, by Wang Laboratories in 1977 and by IBM in 1981. Cell phone technology was operational in 1972, though it would take years for light, affordable cell phones to be generally available. By the mid-1990s the Internet, at first the province of large institutions, became available to every user of a personal computer. Today, billions of individuals walk around with cell phones that have Internet access and more computing power than did IBM's first mainframes.

To say that the health care and medicine have changed spectacularly is an understatement. Recall that the first patent for a medicine, in 1796, was for "bilious pills" that falsely promised relief from a variety of conditions. The only common medical treatment of the day was blood-letting, which probably hastened George Washington's death. Perhaps the first significant medical advance occurred in the 1820s when quinine, from the bark of the cinchona tree, was found to help alleviate the severity of malaria. The germ theory of disease was promoted by Louis Pasteur in the 1860s, and medical knowledge and practice soon accelerated. From the use of x-rays in the 1890s, to the 1920s use of insulin to treat diabetes, to the 1940's introduction of penicillin to treat infections, the advances have been remarkable. In 1955 a vaccine was made available to prevent the dreaded disease of poliomyelitis. Currently, we have vaccinations, organ transplants, CAT Scans, MRIs and, most recently, precision medicine made possible by the mapping of the human genome.

Our knowledge of the Universe, Earth and life have seen at least as meteoric a rise as has our medical knowledge. At the time of the birth of our nation, the age of the Earth was believed to be six thousand years. The concept of life forms existing, evolving and becoming extinct was unknown. Atomic theory was a century away, and it was not even known that life forms are composed of the same stuff as nonlife. It had been revealed that the Earth revolves about the Sun, but the nature of stars and galaxies was a mystery.

It was in the early nineteenth century that the study of organic chemistry was born when Friedrich Wohler showed that living tissue is made from the same molecules as inanimate matter. That the Earth is ancient was discovered at the end of the eighteenth century, and soon after it was revealed that the Earth had been home to creatures long gone. In 1859 Darwin published *On the Origin of Species* and, though there are those who still dispute it today for religious reasons, evolution is the unifying theory of life sciences. Around the same time, Gregor Mendel's experiments set the stage for the science of genetics. Eugene Dubois' discovery of Homo Erectus opened our eyes to our human origins, and Barnum Brown's discovery of the first T. Rex skeleton gave us a glimpse of spectacular creatures who lived tens of millions of years in the past.

By the early twentieth century J.J. Thompson had discovered the electron and Ernest Rutherford had discovered the atomic nucleus. In addition, Albert Einstein turned Classical Physics upside down with his special theory of relativity, showing the equivalence of mass and energy. And his general theory of relativity revolutionized our understanding of spacetime and of gravity. In 1929 Edwin Hubble revealed that the Universe is populated by many "island universes," or galaxies, and that the Universe is expanding. In addition, Niels Bohr, Erwin Schrodinger, Werner Heisenberg and others taught us that at a subatomic level the Universe has a strangeness and randomness that we still have difficulty comprehending. It had always been assumed that the Earth's continents had forever been in their current positions, and that the ocean floors were flat and featureless. Using sonar during World War II, Harry Hess discovered that the ocean floor has deeper trenches and higher mountains than do the continents, and Alfred Wegener's theory of plate tectonics has become the grand unifying theory of Earth science. There has been confirmation of astronomer Fred Hoyle's theory that stars are the cauldrons in which atoms heavier than hydrogen are created.

In 1953 Stanley Miller's experiments showed that organic molecules can be created by conditions that may have been present on the early Earth, and Watson and Crick discovered the double helix structure of DNA. In 1964 Robert Wilson and Arno Penzias detected the cosmic microwave background, corroborating the "big bang" theory of the Universe's origins. Super dense objects peculiarly named "black holes" had been theorized, and in 1971 the first confirmed black hole was discovered. It is now known that supermassive black holes reside in the centers of most, if not all, galaxies. In the latter part of the twentieth century Earth systems science has revealed much about the history of Earth and its systems, and it is now known that human activity, particularly the burning of fossil fuels, is altering Earth's climate. Dozens of unmanned missions have revealed that in our Solar System, the outer planets and their moons are astonishingly diverse, and that conditions favorable to microbial life could possibly exist on Jupiter's moons Europa and Enceladus, on Saturn's moon Titan, as well as in other places. The Hubble Telescope has revealed much more detail about the Universe, and missions such as the Kepler Space Telescope have revealed the existence of planets around other stars. All in all, our knowledge of the Universe has been on an immense upward trajectory.

We have seen that during the 235-year history of the United States, there have been extremely fast and extraordinary advances in transportation, communication, inventions, medicine, and our knowledge of the Universe. And, our government, designed by the founding fathers with checks and balances among the three main branches, is still in place.

However, questions could be asked. How well are we using our governmental institutions to manage our collective lives? Do we do well at advancing our ideas and debating them in an honest, effective fashion? Do we do well at developing better understanding of one another's points of view? Do we do well at resolving our differences and achieving common goals? Do we have a sense of a common destiny? An objective view is that our scientific and technical abilities have far outstripped our ability to cooperate with one another in managing our ever more complex society. That is but one of many perspectives we may have when looking back on our journey through the ten eras of our nation's history. It is my hope that taking this journey has helped my readers gain a more comprehensive grasp of our history with many of its facets. And it is my hope that readers will also come away with a desire to see our human relations skills catch up with our scientific and technical attainments.

References

Allison, Robert J. *American Eras: Development of a Nation 1783 – 1815*. Gale Research, 1997.

Bargeron, Eric. *American Decades 2000 – 2009*. Detroit: Manly, 2011.

Bausum, Ann. *Our Country's Presidents*. Washington, DC: National Geographic Partners, 2017.

Bedore, Pamela. *Great Utopian and Dystopian Works of Literature*. Chantilly, VA: The Teaching Company, 2017.

Blassingame, Wyatt. *The Look-it-up Book of Presidents*. New York: Random House, 2012.

Bolton, John. *The Room Where It Happened: A White House Memoir*. New York: Simon and Schuster, 2020.

Bondi, Victor. *American Decades 1930 – 1939*. Detroit: Gale Research, 1995.

Bunch, Bryan. *The History of Science and Technology*. New York: Houghton Mifflin, 2004.

Bynum, William and Helen, Eds. *Great Discoveries in Medicine*. London, Thames and Hudson, 2011.

Crofton, Ian. *In The Words of The Presidents*. London: Quercus Publishing, 2010.

Davis, Todd and Frey, Marc. *The New Big Book of U.S. Presidents*, Philadelphia: Perseus Books, 2013.

De Bono, Edward, Ed. *Eureka! An Illustrated History of Inventions from the Wheel to the Computer*. New York: Holt, Rinehart and Winston, 1974.

DeGregorio, William A. *The Complete Book of Presidents*. Fort Lee, New Jersey: Barricade Books, 2005.

Dyson, James. *A History of Great Inventions*. New York: Carroll & Graf, 2001.

Freidel, Frank, *Our Country's Presidents*. Washington, DC: National Geographic Society, 1981.

Friedman, Meyer & Friedman, Gerald. *Medicine's Greatest*

Hamilton, Neil A. *Presidents: A Biographical Dictionary*. 3rd Edition. New York: Facts on File, Inc., 2010.

Hazen, Robert M. *Origins of Life*. Chantilly, VA: The Teaching Company, 2005.

Houghton, J.T., Jenkins, G.J., and Ephraums, J.J. *Climate Change: The IPCC Scientific Assessment*, Cambridge University Press, 1991.

Impey, Chris & Henry, Holly. *Dreams of Other Worlds: The Amazing Story of Unmanned Space Exploration*. Princeton, NJ: Princeton University Press, 2013.

Internet Resources
www.history.com
www.medicalnewstoday.com
www.nasa.com
www.science.howstuffworks.com
www.space.com
www.wikipedia.org

Layman, Richard. *American Decades 1950 – 1959*. Detroit: Gale Research, 1994.

Lee, Bandy MD. *The Dangerous Case of Donald Trump: 27 Psychiatrists and Mental Health Experts Assess a President*. New York: St. Martin's Press, 2017.

Logsdon, John, Ed. *The Penguin Book of Outer Space Exploration: NASA and the Incredible Story of Human Spaceflight*. New York: Penguin, 2018.

Launius, Roger. *Space Exploration: From the Ancient World to the Extraterrestrial Future*. Washington, DC: Smithsonian Books, 2018.

McConnell, Tandy. *American Decades 1990 – 1999*. Framingham Hills, MI: Manly Books, 2001.

Nevins, Allan and Commager, Henry. *A Short History of the United States*. New York: Knopf, 1966.

Ness, Erik, "Resistance Fighters: Are Antibiotics Over?" in Brown Alumni Monthly, March/April, 2018.

Stoler, Mark A. *The Skeptic's Guide to American History*. Chantilly, VA: The Teaching Company, 2012.

Trump, Mary L. *Too Much and Never Enough: How My Family Created the World's Most Dangerous Man*. New York: Simon and Schuster, 2020.

Vigliani, Marguerite and Eaton, Gale. *A History of Medicine in 50 Discoveries*. Thomaston, Maine: Phillip House, 2015.

Wiens, Roger. *Red Rover: Inside the Story of Robotic Space Exploration, from Genesis to the Mars Rover Curiosity*. New York: Basic Books, 2013.

Wolff, Michael. *Fire and Fury: Inside the Trump White House*. New York: Henry Holt, 2018.

Woodward, Bob. *Rage*. New York: Simon and Schuster, 2020.

Yenne, Bill. *The Complete Book of US Presidents*. Minneapolis: Zenith Press, 2016.

About the Author

Michael Slavit is a psychologist in private practice. He received his Bachelor's Degree from Brown University, his Master's Degree from the University of Rhode Island and his Doctorate in Counseling Psychology at the University of Texas at Austin.

Though board certified in Behavioral and Cognitive Psychology by the American Board of Professional Psychology, he considers his most important credential to be the confidence of his patients.

Dr. Slavit has a variety of writing interests. He has authored:

>Your Life: An Owner's Guide
>Embracing Fitness
>Train Your Wandering Mind: Coping with ADHD
>Lessons from Desiderata
>Journeys of Imagination: Science Fiction Short Stories
>Cure Your Money Ills

He has works in progress, including:

>225 Limericks
>A Brief History of the Universe, Earth and Life (Illustrated with Limericks)
>My Life as a T Rex

Dr. Slavit is intrigued by how far and fast human civilization has come in the development of science and technology. However, he sees a gap between our science and technology on the one hand, and our philosophical development and human relations ability on the other. He has written, "As a species we have rocketed forward at super speed in developing our science and our technology. But we have lagged so, so far behind in learning to plan for our future, set common goals, resolve our differences, envision a common destiny and uplift one another."

Dr. Slavit hopes that this book is educational and enjoyable, but realizes that some thoughts expressed in the book may be disquieting as well. He believes that being able to face and understand less desirable, as well as more desirable, aspects of our culture and society is important if people are to be motivated to influence society in positive ways, and if human civilization is to make favorable choices.

Made in United States
North Haven, CT
10 April 2023

35290926R00212